MODERN SCIENCE AND AN ANCIENT TEXT

By

Don Kopp

Modern Science and an Ancient Text
Copyright © 1997 by Don Kopp

First Printing 1997

This copy published 2010

ALL RIGHTS RESERVED
This book may not be reproduced in whole or in part
by mimeograph or any other means, without permission.

Library of Congress
Catalog Card Number: 97-93534

ISBN:1453806202
ISBN-13: 9781453806203

Table of Contents

Introduction 3
Forward 5
Ground Rules 9
The Evolution Model 15
The Beneficial Mutation 23
Vestigial Organs 27
The Geologic Column 31
The Fossil Record 35
Ally Oop 47
Survival Of The Fittest??? 53
Evolution The Religion 57
And There Was Light 61
The Creation Model 71
Show Me! 79
God's Terrarium 87
How High's The Water Mama? 99
Glacieogenesis 107
Mount Everest And The Flood Of Noah 115
God's World 121
God's Laws 141
God's Universe 161
Was God An Evolutionist? 167
The World's Greatest Scientists 173
Index 185

INTRODUCTION
MODERN SCIENCE AND AN ANCIENT TEXT

BY: DON KOPP

Before retirement, Don Kopp was the Special Projects Forester for the South Dakota Division of Forestry, where he had been employed for 29 years. During much of his tenure as a fire specialist, Don did research work on fire dependant plant species.

In the realm of fire science and research, Don was able to demonstrate first hand the response and the need of differing plant communities for fire. The idea that differing plants and animals evolved separately is impossible because of the symbiotic relationships demonstrated by the various plant and animal communities. It was this symbiosis in part that made Don, who at one time was a theistic evolutionist, begin to question the validity of evolution as a workable force of nature.

Because of his experience and training in the field of fire and meteorology, Don became the Incident Commander on all large fires for the State of South Dakota and as a member of two National Fire Management Teams, traveled extensively throughout the United States where he was able to see first hand how the various biota in differing regions of the country responded to wildfire as well as prescribed fire.

In 1997, at the request of the Governor's office, Don authored an extensive writing titled: "Prescribed Fire, its Need and the Future." This writing was placed in the State Library as a reference guide for all fire practitioners in the state doing research work on native plant regimes dependent on fire.

After his retirement from the State Division of Forestry, Don started his own forestry consulting business and has been called upon by the Governor's

Office on several occasions to assist in combating several large damaging forest fires and is still actively involved in fire investigation work for various Government agencies and insurance companies.

As an amateur creation scientist and Bible apologist, Don has traveled extensively throughout the United States, the Middle East and Russia, teaching creation science and debating evolutionists whenever the opportunity avails itself. Besides teaching creation science and conducting seminars, he is also well versed in Bible prophecy and apologetics. Don has been able to broaden his knowledge in these fields by involving himself in archaeological digs in Israel as well as attending various lectures by some of the world's leading archeologists.

FORWARD

By L. Ronald Johnson
Professor of Atmospheric Science
South Dakota School of Mines and Technology

As a Christian and long time scientist, I found this book to ask questions that people of science have failed to answer or tried to ignore in the hope that such questions would never be asked. Character assassination of insistent questioners is employed to cover the lack of truth. A diligent research project must be properly guided and managed to keep the search on task and minimize the number of rabbit trails that may lead the searchers astray. The definition of astray in this instance would be any and all distractions that take effort away from the primary focus. What happens when one of these rabbit trails leads to TRUTH?

I dedicated my life to Christ Jesus at a young age and reconfirmed my faith in 2005. Confirmation of my stand for Christ was by water baptism witnessed by the many assembled on the 12th day of March, 2006. I know that Christ Jesus was born of a virgin. He is the Son of the living God. Jesus' shed blood was a sacrifice for my sin and your sin. On the 3rd day Jesus rose from the grave, continued his ministry to His earthly followers, and installed in them the Holy Spirit when He ascended. He now sits on the right hand of our Heavenly Father making ready His return to this world. My admission of sin gave redemption, the Holy Spirit resides in me, and I have been given life eternal as stated by John 3:16.

Scientific research was my primary focus for 31 years when I was a member of various teams of the Institute of Atmospheric Sciences located on the campus of the South Dakota School of Mines and Technology. The Department of Meteorology was established during my tenure, and guiding Master of Science students through their studies and research became a joint effort with our primary research task. Air pollution, Cloud Physics (including x-ray diffraction), Radar Meteorology, and Climatology were primary focus areas. The research generated many journal articles, which were authored or co-authored, many conference papers and presentations, and numerous reports. The Institute of Atmospheric Sciences is self-funded through external contracts and grants primarily with governmental agencies. Service to the campus was through committee assignments given by the President. Near the end of my tenure, I served as the chairman of the Promotion and Tenure Committee; membership of the committee is determined through election by peers.

To receive funding requires peer reviewed proposals that present a novel or unique idea to further scientific understanding. The first hurdle of any proposal is to pass muster with the governmental agency manager. The manager handpicks the reviewers; and, their combined reaction to your proposal, determines a priority level for funding. There are many proposals and only few are funded. Therefore, continued funding requires a reputation of integrity, fiscal responsibility, and a good work ethic. A working relationship with the funding manager is a definite plus to reaping the harvest in this process.

As long as the system of peer review remains truth based, the system works well. But when one considers that the qualified reviewers of your proposal are also competitors for the limited dollars available in that specific program, truth has little chance as worldly lust enters the process. Add to the mix that each of the principle investigators are producing more replications of themselves (i.e. students), and the student's dream is to earn the degree, retain a position with an acknowledged group with a record of many publications, and to continue research in the same area of interest.

The established education system actually adds more pressure for available dollars by producing more qualified candidates. Now consider the consequence when a program manager dictates that fundable work must have a central focus such as global warming. Truth has little chance to survive under the pressures to be successful and maintain a comfortable life style by those doing the research.

Just a closer look at the global warming issue will let us know just how far from the truth the scientific community has strayed. Worldwide temperature data recorded over 100s of years are needed to prove that earth is indeed warming. Surface stations with long term records of more than 100 years are in many populated centers but for rural regions long term records do not exist. Technology has advanced temperature measurement in accuracy, reliability, repeatability, and cost has plummeted. Meteorological surface stations have proliferated as a consequence, but sadly, the long term record is lacking. Many surface stations were established at public facilities. Urbanization of these areas included more and more concrete. Concrete is an efficient solar collector and dissipates the collected energy by radiation. The net effect is labeled heat island. Surface stations established above grassy areas measure temperatures below that of stations over concrete. Many surface stations experienced the heat island effect. Long term temperature trends must avoid contamination of the data by any influence be it heat island and/or technology induced by introduction of a different apparatus to measure the same variable. Another temperature measurement technique is by specific satellites from their positions in space though cloud cover limits what they see. Cloudiness is less of a problem over desert areas and a much larger problem over humid areas where most of earth's population exists.

Once the long term temperature data have been filtered to remove contaminates, the trend for the past century is an increase of almost a half degree. Such a small difference with large deviations for any one year leads to great scientific uncertainty. The alarmist climatologists produced a temperature trend for the same period with a sharp increase over the last decade of the 20th century. This was commonly referred to as the "hockey stick graph" which was based on a very contaminated data set. This graph is the basis for the theory that the earth is warming at an alarming rate and the increased use and dependence on fossil fuels was responsible. The record does indicate that the 1990's were warmer than normal. But, climate does change and that fact is demonstrated by the record of the first decade of the 21st century when average temperatures were below normal. The alarmist climatologists do not recognize this decline. A parody video, titled "Hide the Decline," had more than 500,000 viewers on You Tube and received national attention when Rush Limbaugh played it on his radio program.

A more careful look at carbon dioxide generated by the combustion of fossil fuels in the boundary layer finds that carbon dioxide is not a pollutant because every organism that depends on photosynthesis for life uses carbon dioxide as food. The organism needs the carbon for growth and expels oxygen as a byproduct. The earth's surface is two-thirds water, and water reacts with carbon dioxide to form carbolic acid. Significant increases in carbon dioxide should increase the acidity of earth's bodies of water, but nothing supports the hypothesis. The carbon dioxide budget still remains a mystery.

Can any of our esteemed scientific minds create life or demonstrate that life evolved on a path of many improvements till man appeared? Quite the contrary, true scientific minds of today know there is no mechanism in man's control that produces life, and evolution is impossible. Order cannot come from chaos; but chaos comes from perturbed order. The world is hungry for any theory that will give life without the need for a God. Current world theory supports that intelligent life from somewhere other than earth is responsible for life on earth.

Where is the truth? Jesus said, "For this reason I have come into the world, that I may bear witness to the truth. Everyone who is of the truth hears My voice," John 18:37.

Now why would Jesus say that? This statement tells us that truth is on trial, and truth lacks a defense. Why should truth need to be defended? Ever since the serpent told Eve, "You will surely not die," Genesis 3:4, truth has been under assault by the enemy, and truth is far from us if we are of this world. Further, death did come to the world when Adam and Eve ate the fruit of the tree of knowledge. Truth really doesn't need a defense. We all stand for truth and speak the truth when fitting. Nobody wants to consider that their knowledge may not be truth. Do not confuse truth with justice because justice is a personally reasoned conclusion, which makes it of this world. One word of caution, our sense of truth actually may be the worldview since this is what we see, hear, read, and are taught. The worldview does require a defense even though widely accepted since in reality it is false. How may we know the truth when we hear it? I John 4:4-6, "You are of God, little children, and have overcome them, because He who is in you is greater than he who is in the world. They are of the world. Therefore they speak as of the world, and the world hears them. We are of God. He who knows God hears us: he who is not of God does not hear us. By this we know the spirit of truth and the spirit of error." There is the test plainly stated. If the world hears you, then it is of the world. But, those of God hear the truth.

Are you ready to hear the truth? Prepare yourself by admitting you are a sinner. Anyone in this world duly qualifies and that includes myself. Believe that Christ Jesus came into this world as the Son of God and as a human, died for our sins, and rose from the grave on the 3rd day. Confess our sin to Christ Jesus and ask for forgiveness. Now that you have been made clean, turn the page and learn the truth.

GROUND RULES

Unlike evolution, creation science doesn't contradict the natural laws of the universe. In fact, creation science predicts the existence of the laws which govern the universe.

There are several perceived errors and contradictions in the scripture, and skeptics are quick to point them out. For example, we find this verse in Luke 18:35, while talking about Jesus it says: ***"Then it happened, that as He was coming near Jericho, that a certain blind man sat by the road begging."*** This verse clearly says that Jesus was coming into Jericho.

Then we read the same story as told by Matthew: ***"Now as they departed from Jericho, a great multitude followed Him."*** (Matthew 20:29)

The difference of course is that Luke has Jesus coming into Jericho, while Matthew has Jesus departing. An obvious contradiction; but wait. As in every so-called contradiction, the skeptic has not done his homework. If one were to research this so-called error, he would find instead, a scientific accuracy! Let's take a closer look at the real facts of this story.

The original city of Jericho was destroyed by Joshua in the year 1451 BC, however, Joshua prophesied that the city would someday be rebuilt (Joshua 6:26). This prophecy was fulfilled exactly as Joshua predicted it would be when ***"Heil the Bethelite"*** in the year 930 BC rebuilt the city (I Kings 16:34).

Several hundred years after Jericho was rebuilt, Gentiles began to settle in the same area. However, Jews and Gentiles don't mix, so the Gentiles began to build their own settlement close to old Jericho. This section (a suburb) became known as "New Jericho." Matthew, who was a Jew, had Jesus coming out of Old Jericho, (the Jewish section), while Luke who was a Gentile (and being interested in Gentiles), had Jesus entering the Gentile section. Thus it can be ascertained (by reading both Gospels) that Jesus met the blind man half way between Old Jericho and New Jericho, and Mark (a third Gospel) tells us who that blind man was! (Mark 10:46)

There are also several places where scientific error can be shown where scientific accuracy is not intended. Figurative or culturally common expressions are not always to be identified as scientific or unscientific. An example of this is the Bible speaking of a "sunrise" or the "sun-setting." Even today we use this unscientific term, but it is a word which accurately describes an event (the sun does rise and sets by virtue of the earth's rotation) and a literal scientific application is not required. But just to keep the record straight, Job 38:14 tells us that the earth is turning, a process which ultimately causes the sun to rise or to set.

In Genesis 7:11 we find the term ***"windows of heaven"*** and from this statement alone, we could probably surmise an array of inaccurate babblings and point a skeptical finger at a text filled with mythology akin to the Babylonian account of "sluice gates" in heaven made of solid wooden beams. It was from these "sluice gates," according to the Gilgamesh epoch of the flood that the water poured out upon the earth and the world and its inhabitants perished.

The Biblical term ***"windows of heaven"*** are descriptive of another source of water from the heavens other than clouds. It was from here the windows of heaven along with several other sources that water spilled forth upon the earth, ultimately covering the entire planet and destroying all life with the exception of those safely aboard the ark which was built by Noah and his family.

We will be covering the scientific accuracy of these verses and others related to the flood in a later chapter.

Should one look through the pages of the Hindu bible he would find this book teaches that the world is held up on the back of an elephant, the elephant is in turn standing on a giant turtle which is swimming in a cosmic sea. The Greeks believed that Atlas stood holding the earth on his neck and shoulder.

The Egyptians believed the earth started out as a winged egg and flew around in space until it hatched.

If a person were to do a study of the books just mentioned, one would find that they promote the science or beliefs of their day; albeit a false science. Incidentally, if we turn to the Scripture and study what holds the world up, we would read: *"He* (God) **spreads out the heavens over empty space, and He hangs the earth upon nothing."** (Job 26:7)

There are some basic laws which God established that govern the universe. These laws will not change of their own volition and cannot be transgressed in the natural. It is these laws that make up the ground rules around which science must function. Below are several basic scientific laws which any theory, in order to be considered valid, should <u>always</u> function within.

Matter: Matter and energy are different manifestations of the same thing. Everything in the cosmos is made up of matter; it cannot be created (it is impossible for nothing to produce something,) nor can matter be destroyed. **This is the first law of thermodynamics.**

Second Law of Thermodynamics: This law states: Everything that exists is matter (or energy), everything that happens is energy conversion. Once this energy is expanded it cannot be recalled to do the same job again, thus the universe as a whole, is running down like a wound up clock. Isaac Asimov said of this law: *"As far as we know all changes are in the direction of increasing entropy, of increasing disorder, of increasing randomness, of running down."* [1]

By far the most common response by evolutionists to the problem encountered by the second law is to deny its applicability to an open system such as the earth. Since there is enough energy reaching the earth from the sun to more than offset the loss of energy in its processes due to entropy, they say, the problem is irrelevant. However, this response is itself irrelevant since it confuses quantity of energy of which there is certainly enough, with conversion of energy. The question is not whether there is enough energy from the sun to sustain the evolutionary process; the question is how does the sun's energy sustain evolution?

Forces and Fields: Interactions in nature depend upon three types of force and the fields associated with them; namely, electromagnetic, gravitational and nuclear forces. All three have apparently always acted as they do now since the beginning of the universe.

I have briefly outlined several of the more empirical laws of our universe. A good scientific theory should work within the framework of these laws. Webster's dictionary describes a theory this way: *"The general principles on which a science is based and built up."*

Should I postulate a theory which would indicate that all loose rocks roll uphill, my postulate would be in violation of the laws of gravity and my theory would, of course, be wrong.

If there are no known laws upon which to build a theory, then the theory becomes a religion. A belief that requires the universe to come into existence from nothing is based upon faith, not science! Colin Patterson, senior paleontologist at The British Museum of Natural History said in a talk he gave at The American Museum of Natural History, November 5, 1981, that he realizes that in accepting evolution he had moved from science into faith.

In a recent BBC program, Dr. Patterson stated that: *"...all we really have of the evolutionary phylogenetic tree are the tips of the branches. All else the filling in of the trunk and of the branches is simply story telling of one kind or other."*

Then, evolution as you will see, does not fit the known laws which govern the universe, it contradicts those laws. Creation science on the other hand, not only functions normally within the universal laws, it predicts those laws.

Stansfield, an evolutionist scientist and professor at California Polytechnic Institute, describes the laws of thermodynamics this way: *"The First Law of Thermodynamics, sometimes called the Law of Conservation of Energy, states that energy can be transformed from one kind to another, but it can neither be created nor destroyed. Since matter and energy have been shown to be interconvertible, the First Law can be modified to state that neither matter nor energy can be created or destroyed. The Second Law of thermodynamics states that in converting one form of energy to another, some of it is lost as unusable heat. Entropy is the thermodynamic quality of randomness or disorder within a system. The Second Law therefore implies that as energy is being transformed throughout the universe, entropy is increasing. These laws argue strongly for a created universe."* [2]

References

1. Isaac Asimov, "In the Game of Energy and Thermodynamics, You Can't Even Break Even"
2. W. Stansfield, "The Science of Evolution" p.57

THE EVOLUTION MODEL

Because evolution is taught in our country today as if it were a proven fact, many Christians and entire church organizations mix evolution with Christian ideology. The idea that God used evolution to bring about His purpose is known as "theistic evolution." It is causing a departure in the acceptance of the Scripture as literal, inerrant truth. This rejection of God's Word as infallible can only lead to apostasy, and finally bondage. Duane Gish, eminent biologist and creationist wrote this about the church accepting evolution: *"The acceptance of the theory of evolution has promoted apostasy because it has caused a radical change in the view of Scripture."* [1]

History has shown that the church and/or nation that eventually fall away from a total acceptance in God's Word as literal truth slowly sink into ignorance and finally bondage. The first signs of a people starting to falter is an increase in ignorance or a decay in moral philosophy. We write books in the name of education with titles like: "The Population Bomb," in which the author (an avid evolutionist), describes pregnancy as *"a nine month disease."* [2]

Recently a scientist from the Institute For Creation Research was visiting with a clergyman from the former Soviet Union. He asked the clergyman if the teaching of communism in the public schools led their nation into bondage. The clergyman replied: *"No, the thing that led the nation into utter bondage by the Communists was when the school system replaced science with the teaching of evolution."*

Karl Marx dedicated his book Das Kapital to Charles Darwin. Since Darwin (the propagator of evolution) and Marx (the father of communism), one half of the world has been led into enslavement! The Bible says: **"Therefore My people have been led into captivity because they have no knowledge."** (Isaiah 5:13) As shown above, the teaching and acceptance of evolution is a regression of knowledge. The acceptance by our society of abortion and homosexuality are two more examples.

Again, quoting Colin Patterson, senior paleontologist at the British Museum of Natural History, in a speech before the American Museum of Natural History on November 5, 1981, he said this of evolution: *"The theme is that evolution not only conveys no knowledge but it seems somehow to convey anti-knowledge."* And along a similar vein, J. Cracraft concurs: *"....axiomatic reasoning, as it is employed by functional evolutionary morphology, simply does not increase our knowledge about process"* [3]

The Bible says: **"Therefore people who do not understand will be trampled."** (Hosea 4:14) We need to come to an understanding that God is the Creator and His Word is infallible.

It is at this point that I should give you the meaning of the word evolution so as to avoid any misunderstanding.

When I use the term "evolution," I am referring to a naturally occurring beneficial change which produces increasing complexity. When referring to the evolution of life, this increasing complexity would be shown if the offspring of one form of life had a different, improved and reproducible set of vital organs that its ancestors did not have. This is sometimes called organic evolution, the molecules-to-man theory or macroevolution.

Microevolution, on the other hand involves only such changes as different shapes, colors, sizes, or minor chemical alterations. Changes which both creationists and evolutionists agree are relatively trivial and easily observed. It is macroevolution then, that is being so hotly contested today and this is the meaning of the term evolution.

There is no proof, nor is there any scientific test by which evolution can be shown to have happened. Professor David Allbrook, Professor of anatomy at the University of Western Australia, said that *"evolution is a time-honored tenet of faith."*

The Big Bang

Depending on which dating method you use, the universe was blown into existence ten to twenty billion years ago. Of course one of the first questions a person should ask is: "Where did the matter come from that made up the material that exploded?" Is this idea consistent with the first law of thermo-dynamics?

"The conservation of energy, one of the most cherished principles of physics, is violated in the big bang model...Thus it is impossible for matter to come into existence without violating energy conservation. It is customary to water down this difficulty by statements like 'the laws of physics break down at a singularity;' however the essential truth remains the same." [4]

This problem of where the material came from that made up the big bang has left some scientists highly unsatisfied with the theory: *"A number of scientists are unhappy with the big bang theory....For one thing, it leaves unanswered the questions that always arise when a precise date is given for the creation of the universe: 'Where did the matter come from in the first place?'"* [5]

The next question a person should ask is: "Do explosions produce order?" Is this theory consistent with the second law of thermodynamics? Of course the answer to both questions is absolutely not! Random processes do not arrange themselves into order but rather disorder. According to the first law, you cannot get something from nothing. The second law requires that ultimately all energy changes tend to be downward, not upward, as evolution would require. When a builder constructs a house he doesn't stack all his building materials on a keg of powder and blow his home into existence. This of course, would require a random process (the explosion) to produce order (the house).

One of the leading cosmologists in the world, Fred Hoyle writes," *The big bang theory holds that the universe began with a single explosion. Yet as can be seen below, an explosion merely throws matter apart, while the big bang has mysteriously produced the opposite effect, with matter clumping together in the form of galaxies."* [6]

Bible science would dictate that a house needs a builder, in fact, Proverbs 24:3 says: **"Through wisdom a house is built, and by understanding it is established."**

Let us assume, for a moment that the big bang randomly occurred. Thus we would expect that the materials which make up the universe should be made of the same elements because it all came from the same place didn't it? So what does the study of astronomy tell us about the molecular make up of the observable universe? Astronomer Donald H. Menzel tells us: *"The material of the Earth and Mars, Venus and Mercury should all be hydrogen and* helium, similar to that of the sun and the rest of the visible universe: Actually much less than 1% of the Earth's mass is hydrogen or helium."[7]

Actual scientific study reveals that the stars are different one from the other as well as the planets. Bible science already made that prediction two thousand years ago: **"There are celestial bodies and terrestrial bodies; but the glory of the celestial is one, and the glory of the terrestrial is another. There is one glory of the sun, another glory of the moon, and another glory of the stars; for one star differs from another star in glory."** (I Corinthians 15:41, 42)

As mentioned before, the complex order of the universe argues against a random explosion producing it. One of the greatest scientists of all time, Sir Isaac Newton, said this of the solar system: *"The six primary planets are revolved about the sun in circles concentric with the sun, and with motions directed towards the same parts, and almost in the same plane. Ten moons are revolved about the earth, Jupiter, and Saturn, in circles concentric with them, with the same direction of motion, and nearly in the planes of the orbits of those planets; but it is not to be conceived that mere mechanical causes could give birth to so many regular motions, since the comets range over all parts of the heavens in very eccentric orbits."*[8]

Robert Jastrow, founder and director of NASA"s Goddard Institute for Space Studies, in his analysis of the abundance of unburned hydrogen in the cosmos, deduces that it almost certainly requires a *"creation"* of the universe. Likewise, VanWylen, Chairman of the Department of Mechanical Engineering at the University of Michigan, in his textbook on thermodynamics also supports the idea of "creation" as the only viable answer to the problems presented by the laws of thermodynamics.

Astronomer, Sir Fred Hoyle of Great Britain, who is referred to as an *"eminent cosmologist"* by Steven J. Gould of Harvard, recently made this comment about the big bang: *"As a result of all this, the main efforts of investigators have been in papering over holes in the big bang theory, to build up an idea that has become ever more complex and cumbersome...I have little hesitation in saying that a sickly pall now hangs over the big bang theory. When a pattern of facts become set against a theory, experience shows that the theory rarely recovers."*[9]

Let's give evolution the benefit of doubt and assume that the complexity of the atom came from nothing and built itself into a ball about three feet across. According to evolutionists, this is how it all began; the entire universe was blown into existence from this small ball. This tiny chunk of mass caused the greatest explosion of all time and the universe was born out of this incomprehensible blast. Therefore, we must assume that nothing decided to create itself, and this creation became the universe which became god to many. Evolutionist and cosmologist, Dietrick Thomsen, writes this enlightening, highly scientific statement: *"The universe wants to be known. Did the universe come about to play its role to empty benches?"* [10]

The answer to that question of course (according to Thomsen), is "no." The universe didn't create itself before an "empty bench," it also created you and I, so we could watch it do its thing, thus making the universe and its creation god.

Perhaps G. Burbidge summed it up best: *"Probably the strongest argument against a big bang is that when we come to the universe in total and the large number of complex condensed objects in it, the theory is able to explain little."* [11]

Molecules to Man

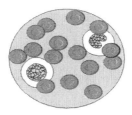

The next question one should ask himself is, how could that first life form develop on a sterile ball? One should also ask this question, where did all the water come from that life supposedly got its start in?

Things are beginning to get complex now. We just went from nothing to mass, from mass to an explosion, from an explosion to a place called Earth where different elements are combining to create water, hydrogen, nitrogen, helium, and all the other elements that made up the primeval Earth. Is this contradictory of the observable laws? Now this wondrous "singularity" (one time event) happened all by itself at the direction of the ......universe??? Let's take a quick look at what real scientific experimentation and observation reveals.

Dr. Duane Gish, noted biologist, had this to say concerning the random processes that supposedly created life: *"If the earth, early in its alleged evolution, had oxygen in its atmosphere, the chemicals needed for life would have* been removed by oxidation. But if there had been no oxygen, then there would have been no ozone, and without ozone all life would be quickly destroyed by the sun's ultra violet radiation." [12]

P. Punn, Professor of biology at Wheaton University, concurs with Gish and his findings: *"Under the primordial conditions, many experiments designed to synthesize bio-organic molecules failed when molecular oxygen was present; however, they, succeeded when oxygen was removed."* [13]

Continuing along this same line of thought, we find more proof of the degradation of the basic blocks for life by sunlight and/or oxygen. *"In order for the primordial polypeptides, polysaccharides and polynucleotides to accumulate in any amount in the primordial sea or in localized aqueous systems, the rate of their formation must have exceeded the rate of their degradation. However, this stipulation is contrary to what is expected from the second law of thermodynamics. Therefore a serious paradox exists in stage two of chemical evolution and cannot be easily ignored."* [14]

There is also a problem in the formation of the amino acid sequence. This is addressed by author Michael Denton who holds a Ph.D. in molecular biology. Denton writes: *"The really significant finding that comes to light from comparing the proteins amino acid sequences is that it is impossible to arrange them in any sort of evolutionary series."* [15]

Thus the evolutionist must assume that a primordial soup (whether in the form of seas or evaporating little pools) existed in which would be gathered the essential elements for the possible formation of amino acids, proteins, and the genetic coding system, protocells and other evolutionary stages. Charles Darwin said of this so-called primordial pond: *"But if (and oh! What a big if!) we could conceive in some warm little pond, with all sorts of ammonia and phosphoric salts, light, heat electricity etc., present, that a protein compound was chemically formed ready to undergo still more complex changes."* [16]

So what are the scientific possibilities of Mr. Darwin's *"warm little pond"* where all the right elements gathered, and life spontaneously began? D. Hull writes: *"The physical chemist, guided by the proved principles of chemical thermodynamics and kinetics, cannot offer any encouragement to the biochemist, who needs an ocean full of organic compounds to form even lifeless coacervates."* [17]

Evolutionist and biochemist Sydney Harris bemoans this same problem: *"How can the forces of biological development and the forces of physical degeneration be operating at cross purposes? It would take, of course, a far greater mind than mine even to attempt to penetrate this riddle. I can only pose the question."* [18]

What about our *"primordial atmosphere"*? R. Shapiro, professor of chemistry at New York University: *"We have reached a situation where a theory has been accepted as fact by some, and possible contrary evidence is shunted aside. This condition, of course, again describes mythology rather than science."* [19]

Scott writes in The New Scientist: *"How much of this neat tale is firmly established, and how much remains hopeful speculation? In truth, the mechanism of almost every major step, from chemical precursors up to the first recognizable cells, is the subject of either controversy or complete bewilderment."* [20]

Probably, W.R. Bird sums it up the best, as he points out the impossibility of a primordial atmosphere producing any kind of life: *"The oxygen-free early atmosphere that is necessary to biochemical evolution is more a matter of wishful thinking than of scientific fact, and strong scientific evidence directly contradicts the possibility of such an atmosphere, according to a number of evolutionist scientists. There is no evidence for...but much against a primitive oxygen-free and methane-ammonia-rich atmosphere."* [21]

Louis Pasteur did more for medical science and the health of the world than anyone before or since. Although he was a notably insightful scientist, any others would have had as much opportunity of doing what he did if they had started out with the Biblical presupposition that things reproduce only *"after their kind."* It is very difficult to stop a pathogenic bug from spreading if you think it could spontaneously pop up in your body. The faith that is required to believe this way is truly amazing. It is akin to a Christian saying that he believed God to be an utter impossibility. Such a person would be branded as a mindless fanatic by society. Yet if one makes this statement in the name of science....no problem!

But if one reasons, as the creation model shows, that for every disease organism there is a parent and a grandparent, and so on, then disease pathways can be sought, traced, and blocked.

It can safely be said in this writing and without contradiction, that spontaneous generation has <u>never</u> been observed. Life as far as we know only comes from life. No man has ever observed life come into existence in any other way than in the Biblical way. All the life forms that are on this planet now, have always been here from the beginning. The smartest man who ever walked the planet wrote: *"There is nothing new under the sun."* (Ecclesiastes 1:9)

Authors Note:

As of this writing, Dr. George Wald, Nobel Prize winner of Harvard University, made this statement concerning the origin of life on our planet: *"One has only to contemplate the magnitude of this task to concede that the spontaneous generation of living organism is impossible. Yet here we are - as a result."*

References

1. Duane Gish, "Evidence Against Evolution p.19
2. Paul Ehrlich, "The Population Bomb"
3. Carcraft, "The Use of Functional and Adaptive Criteria in Phylogenetic Systems," American Zoologist, (1981) Narlikar & Padmanabhan, "Creation-Field Cosmology: A Possible solution to singularity, horizon, and flatness problems," p.32 (1985)
4. A. Krauskopf & A. Beiser, "The Physical Universe" p.645 (1983)
5. F. Hoyle, "The Intelligent Universe" p.184, 185 (1983)
6. Donald Menzel, "Astronomy" p.21
7. Isaac Newton, "Mathematical Principals" p.543-544
8. Fred Hoyle, "The Big Bang Under Attack" Science Digest, Vol. 92, p.84
9. D. Thomson, "A Knowing Universe Wants To Be Known" p.124
10. G. Burbidge, "Was there really a Big Bang?" Nature, p.36 (1971)
11. D.Gish "Speculations and Experiments Related to the Theories on the Origin of Life"
12. P.Punn, "Evolution" p.8
13. A.L Laehninger, "Biochemistry" (2nd edition), p.1038
14. M. Denton, "Evolution: A Theory In Crises" p.289
15. C. Darwin, "The Life and Letters of Charles Darwin" p.202
16. Hull, "Thermodynamics and kinetics of spontaneous generation" p.186, Nature 693, 694
17. S.Harris, "Second Law Of Thermodynamics" The San Francisco Examiner, Jan. 1984
18. R. Shapiro, "Origins: A Skeptic's Guide to the Creation of Life on Earth" p.12
19. Scott, "Update on Genesis" New Scientist, p.30
20. Bird, "The Origin of Species Revisited" p.333

THE BENEFICIAL MUTATION

After life supposedly started from a simple cell, it is theorized that it began to mutate upward. That is, its process of becoming more complex was brought about by slight changes in the genetic code of the cell. These slight changes or variations would have to be vertical in order to be of any value to its progress toward becoming a higher life form. These genetic variations, it is believed, took place over millennia. Evolutionists call these changes "horizontal variations." Does that sound confusing to you? If it doesn't it should, because, as the term implies, "horizontal" is really neutral, or level, it does not go up as it must in order for the cell to progress toward becoming more complex. Thus, horizontal variations don't really mean anything. *"Horizontal variations are not real evolution, of course, nor are mutations, which are always either neutral or harmful, as far as all known mutations are concerned. A process which has never been observed to occur in all human history should not be called scientific."* [1]

No known mutation has ever produced a form of life having greater complexity and greater viability than its ancestors. There is also no reason to believe that mutations could ever produce any new organs such as the eye, the ear, or the brain.

While there is no question that mutations are occurring today, and have done so in the past, there is no scientific evidence to support the idea that mutations are beneficial at all. In fact, Pierre Grasse, who is himself an evolutionist, wrote: *"No matter how numerous they may be, mutations do not produce any kind of evolution."* [2]

Not only do mutations not produce *"any kind of evolution,"* random mutations are in actuality harmful to the organism in which it is occurring. *"The Darwinian theory is wrong because random variations tend to worsen performance, as indeed common sense suggests they must do."* [3]

Evolutionists also believe that a process called natural selection has produced evolution. Natural selection however is a tautology, that is, it explains nothing. Evolutionists like to use the peppered moth as their shining example of evolution in progress through natural selection. The peppered moth was a white colored moth found in Europe.

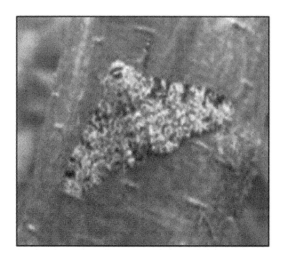

The moth lived on white barked trees growing there. During the industrial revolution which began in the 1700s, coal fired plants began to turn the white barked trees a dark gray because of the air pollution. The birds which couldn't easily identify the white moths against the white barked trees, could now easily spot them, and began to gobble them up. Over a period of years the darker colored moths became the dominant ones while the white moths became quite rare.

This, according to evolutionists, is a beautiful example of evolution in progress through natural selection. The question one should ask himself is this: did the darker colored moths survive because of natural selection? The answer to the question obviously is yes. Why did they survive? Because they were dark colored! Why were they dark colored? Because they survived!

That is a tautology. It tells us nothing about evolution because in the end, it's still a peppered moth. In his book titled, "Introduction to Darwin", L. Matthews, a staunch evolutionist concedes: *"....they do not show evolution in progress, for, however the populations may alter in their content of light, intermediate or dark forms, all the moths remain from beginning to end Biston betularia."* [4] They will remain forever, peppered moths.

Many other experiments have been done using fruit flies to try and show evolution. In the experiments fruit flies were subjected to high doses of radiation so as to cause various forms of mutations. Fruit flies were chosen because they are prolific breeders and thousands can be produced from one set of parents in one season. It was hoped that by consistently breeding the abnormal ones that a new species of fly would appear over a relatively short period of time, thus the process of evolution could be shown, at least to some degree.

While the experimenters were able to change the color and in some case the size and shape of the flies, the end results were always the same: *"Over 70 years of fruit fly experiments, equivalent to 2,700 human generations, give no basis for believing that any natural or artificial process can cause an increase in complexity or viability. No clear genetic improvement has been observed despite the many unnatural efforts to increase mutation rates."* [5]

In his book titled "Scientific Creationism" author Henry Morris writes: *"Even if the mutations are not harmful enough to cause their carriers to be eliminated completely by natural selection, the over-all effect is to gradually lower the viability of the population."* [6]

Once again, if we test our scientific theories such as a beneficial mutation against the known laws of our planet (the second law of thermodynamics says that random processes are downward), then we can only conclude that mutations are harmful, and at best, neutral. This is exactly what the creation model would predict and so (it would seem), that creation science would be accepted as the superior science.

References

1. David Kitts, "Paleontology and Evolution Theory" Vol. 28, p.466
2. P. Grasse, "The Evolution of Living Organisms" p.87, 88
3. F. Hoyle, "The Intelligent Universe: A New View of Creation and Evolution" p.48
4. Mathews, "Introduction, to C. Darwin, The Origin of Species" p.11 (1971)
5. Monroe Strickberger, "Genetics" (2nd edition), p.812
6. Henry Morris, "Scientific Creationism" (Public School edition) p.56

Notes

VESTIGIAL ORGANS

Evolutionists have for a long time clung to the idea that certain organs of the body were the remnants of our evolutionary past. These organs such as the appendix, while useful millions of years ago, have no apparent use today and so they present an argument in favor of evolution. After all, if there really was a God He surely wouldn't create man with organs that had no function now would He? Webster's dictionary has this to say about the appendix: *"A small closed tube growing out of the large intestine. It serves no known purpose."* Other so-called vestigial organs are the tonsils, the thyroid gland, ear muscles, the coccyx, the pineal gland and the thymus, to name a few.

As a boy I can still picture the cave man in my high school text book standing in front of his cave home with a large piece of raw meat in his hand. The story is told how early man needed his appendix back in those days to help him digest raw meat as he had not yet learned how to build a cooking fire. Thus, the myth is perpetuated by the use of pictures in school textbooks which show this brutish, sub-human animal chewing on a piece of raw meat. I can still remember thinking to myself; gee, it's no wonder I like raw hamburger!

That's the basic scenario and the reasoning behind vestigial organs. Dr. Henry Morris writes: *"In view of the history of this subject, it would seem the better part of wisdom not to claim any organs at all as vestigial. The ignorance of scientists about the specific functions of such structures does not prove they have none. It is more likely than not that in the very few cases remaining more intensive study will, as it often has in the past, reveal specific functions actually accomplished by these supposedly useless organs."*[1]

The existence of human organs whose function is unknown does not imply that they are vestiges of organs from our evolutionary ancestors. In fact, as medical knowledge has increased the function of almost all of these organs has been discovered.

"The status of other structures formerly thought to be vestigial is now being challenged, including human male nipples, tonsils, and certain portions of the whale skeleton. The interpretation that the nipples in the human male are inherited from an ancestor in which they were functional is being questioned. The tonsils were once considered useless but are now known as part of the lymphoid mass that traps infectious agents." [2]

Likewise, evolutionist S. Scadding summarizes the research he did along these lines: *"Since it is not possible to unambiguously identify useless structures, and since the structure of the argument used is not scientifically valid, I conclude that 'vestigial organs' provide no special evidence for the theory of evolution."* [3] And, *".... I would suggest that Wiedersheim was largely in error in compiling his long list of vestigial organs. Most of them do have at least a minor function at some point in life."* [4]

Another type of organ that was believed to show the different stages of evolution is also a myth and still being perpetrated in our public class rooms today. This is a myth known as "recapitulating embryos." This idea was dreamed up and promoted by a man named Haeckel, who intentionally made some fake drawings of differing animal and human embryos with the intention of showing that all embryos proceed through the same evolutionary stages. Haeckel dressed up his lie with a fancy scientific sounding name; "The *Biogenetic Law."* This little bit of absurdity is still held as fact by many evolutionists who either don't know any better, or are dishonest. W. Bock, who was a biology professor at Columbia: *"....the biogenetic law has become so deeply rooted in biological thought that it cannot be weeded out in spite of its having been demonstrated to be wrong by numerous subsequent scholars. Even today both subtle and overt uses of the biogenetic law are frequently encountered in the general biological literature as well as in more specialized evolutionary and systematic studies."* [5]

Likewise, Mayr of Harvard describes the biogenetic law as invalid. *"The embryo in its early formation looks as if it has a tail stage of development. The embryos of certain animals had tails and it is taught that these tails grow into legs."* [6] Strickburger: *"As an embryo develops, it does not pass through the adult stages of its alleged evolutionary ancestors. Embryologists no longer consider the superficial similarity that exists between a few embryos and the adult forms of simpler animals as evidence for evolution."* [7]

In summing up this portion of "vestigial organs," I quote Dr. Harry Rimmer: *"There is no animal known to man whose tail develops into legs and the contention that this is true in the embryo is a pitiable evidence of the weakness of the case for evolution."* [8]

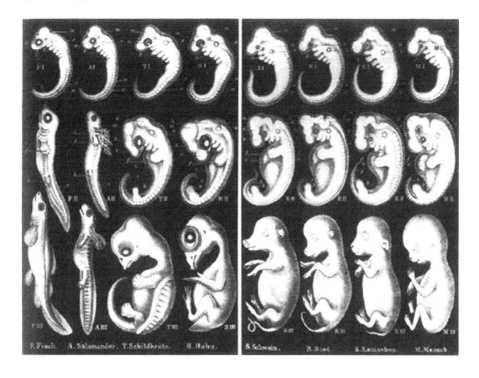

Haeckel's phony drawings

References

1. Henry Morris, "Scientific Creationism" p.76 (Public School ed.)
2. J.T. Barrett, M.D., "Textbook of Immunology" p.6 (2nd ed.)
3. Scadding, "Do 'Vestigial Organs' Provide Evidence for Evolution?" (from the book Evolution Theory)
4. Rager, "Human Embryology and the Law of Biogenesis", (Biology Forum) p.456
5. Bock, "Book Review, (Science) p.164
6. E. Mayr, "The Growth of Biological Thought" p.215
7. M. Strickburger, "Genetics" p.812 (2nd ed.)
8. H. Rimmer, "The Theory of Evolution and the Facts of Science" p.65

Notes

THE GEOLOGIC COLUMN

It is held as fact and taught in our schools that vast periods of time were required to form the earth's crust. The different layers of rock that you can see exposed in various parts of the world were supposedly laid down, layer after layer over millions of years. The Scottish geologist James Hutton maintained that *"the present is a key to the past,"* in other-words, events we observe to be happening on earth today, were the exact same events which occurred millions of years ago.

Further, it is these observable events that can account for all the geologic features we see in the column. This philosophy became known as the doctrine of uniformitarianism. It is called this because it demands extremely long periods of time, where earth changes have been slow and uniform and these changes have occurred so slowly and uniformly that they escape notice.

Ironically, Peter, one of Jesus' disciple's prophesied that this exact doctrine would become a latter day teaching; he wrote, **"Knowing this first: that scoffers will come in the last days, walking according to their own lusts, and saying, 'where is the promise of His coming? For since the fathers fell asleep, <u>all things continue as they were from the beginning of creation</u>.' For this they willfully forget: that by the Word of God the heavens were of old, and the earth standing out of water and in the water, perished, being flooded with water."**
(II Peter 3:3-6)

Scripture claims the earth's crust as we know it today was formed through flooding and in a relatively short period of time, not, as evolutionists believe, gradually and uniformly over hundreds of millions of years. In truth, the idea of a geologic column was first conceived in the *1800's* by creation scientists as W. Bird correctly states: *"The stratigraphic column and our general understanding were thus developed empirically and without any reference to evolutionary assumption by scientists who nearly all found the geological record consistent with the theory of creation and non-supportive of the theory of evolution."* [1]

Since schools and colleges still teach the geologic column, and geologic uniformity, the first question you should ask is: "How is the age for each layer of the column determined?"

The answer to that question is quite simple. Each successive layer of rock is dated by the fossils found within that layer. These fossils are called "index fossils." Sounds like a reasonable way to determine the age of a certain layer of rock doesn't it?

It is at this juncture that you should ask the next question: "If the fossil dates the rock, how is the age of the fossil determined?" The answer to that question is: "By the rock the fossil is found in." At this point you're probably thinking to yourself that there must be some mistake. How can the fossil date the rock, and the rock date the fossil? Isn't this circular reasoning?

The Geology Encyclopedia Britannica has this to say about this method of dating: *"It cannot be denied that from a strictly philosophical standpoint geologists are here arguing in a circle. The succession of organisms has been determined by a study of their remains embedded in the rocks and the relative ages of the rocks are determined by the remains of organism they contain."* (Volume 10, p.168)

This type of circular reasoning is also questioned by O'Rourke who writes: *"The intelligent layman has long suspected circular reasoning in the use of rocks to date fossils and fossils to date rocks. The geologist has never bothered to think of a good reply, feeling the explanations are not worth the trouble as long as the work brings results."* [2]

As of late, the idea of geological uniformity is being questioned more and more by some evolutionists. Steven J. Gould: *"Substantive uniformitarianism as a descriptive theory has not yet withstood the test of new data and can no longer be maintained in any strict manner."* [3]

Gould again writes: *"Often, I'm afraid, the subject is taught superficially, as with Geikie's maxim 'the present is the key to the past' used as a catechism and the smokescreen to hide confusion both of student and teacher."* [4]

As I have just demonstrated, dating of the geologic column and the idea of uniformitarian geology isn't really science at all, but rather a poorly thought-out idea that is foisted upon an unsuspecting public in the name of science. The promoting of the geologic column however is good science as long as each layer isn't interpreted as taking millions of years for its formation.

THE GEOLOGIC COLUMN

RECENT
PLEISTOCENE - 975,000 YEARS
TERTIARY - 79 MILLION YEARS
MOSOZOIC - 200 MILLION YEARS
PALEOZOIC - 300 MILLION YEARS
PROTEROZOIC - 1,000 MILLION YEARS
ARCHAEOZOIC - 1,800 MILLION YEARS

References

1. W. Bird, "The Origin of the Species Revisited' Vol. 1, p.185
2. J.E. O'Rourke, "Pragmatism Verses Materialism in Stratigraphy" American Journal of Science, Vol. 276, p.48
3. S.J. Gould, "Is Uniformitarianism Necessary?" American Journal of Science, Vol. 263, p.263
4. S.J. Gould, "Is Uniformitarianism Useful?" Journal of Geological Ed. Vol. 15 p.24

THE FOSSIL RECORD

"Fossils are the remains of pre- historic life or some other direct evidence that such life existed. To become fossilized a plant or animal must usually have hard parts, such as bone, shell, or wood. It must be buried quickly to prevent decay and must be undisturbed throughout the long process. Because of all this very few plants or animals that die are preserved as fossils.[1]

The fossil record is the pride and joy of all evolutionists. It is here that claims are made that one can present a good case for the theory of evolution. It is also here that evolution stands as a shining tribute to all the world's humanists and atheists. So excited was Karl Marx (the father of communism) over Darwin's ideas about evolution, that Marx dedicated his book Das Kapital to Charles Darwin. Evolution, the religion of atheists, took Russia by storm.

A question one should ask himself of the fossil record: "does it really support evolution?" One of the most distinguished of all French Zoologists said of the fossil record: *"Naturalists must remember that the process of evolution is revealed only through fossil forms."*[2] Noted biologist, C.O. Dunbar writes in agreement: *"Fossils provide the only historical, documentary evidence that life has evolved from simpler to more complex forms."*[3]

Finally, thinks the fledgling evolutionist, we are beginning to get some real evidence for our brand of science; the "ole" fossil record will show those religious fanatics once and for all.

If evolution is true, and the fossil record documents the process well enough for evolutionists to make such bold claims, then we should be able to see from the fossil record simple life forms in the lower, older rock formations. Further, we should be able to see a steady increase in life's progress from those simple forms to a higher, more complex state. These life forms would culminate at the top of the column with the remains of Homo erectus.

Moreover, there should also be found an abundance of transitional fossils, that is, fossil remains of animals that were in various stages of evolution, or the "link" between differing species. *"As living forms diverged into the millions of species which have existed in the past and which exist today, we would expect to find a slow and gradual transition of one form into another."*[4]

Keeping these things in mind, let us turn to the fossil record which *"provide the only historical, documentary evidence"*[5] for evolution and what do we find?

"The known fossil record fails to document a single example of phyletic evolution accomplishing a major morpho-logic transition."[6]

One of the leading proponents of evolution in our nation today, Stephen J. Gould (Gould is an admitted Marxist), said this about the fossil record: *"I regard the failure to find a clear vector of progress in life's history as the most puzzling fact of the fossil record. We have sought to impose a pattern that we hoped to find on a world that does not really display it."*[7]

Along a similar vein, we read: *"....fossil species remain unchanged throughout most of their history and the record fails to contain a single example of a significant transition.*[8]

The following pictures are of actual fossils found as they first appear in the fossil record. There are no predecessors in the record leading up to them, nor are there any changes from their modern day counterparts.

Common bass from the Green River formation, Wyoming

Eocene Spider and 2 Wasp – 54 million years old

Wasps (Hymenoptera) and Spider (Araneae)

The Fossil Record

5 ft. tall palm frond found in the Green River formation (55 million years ago!)

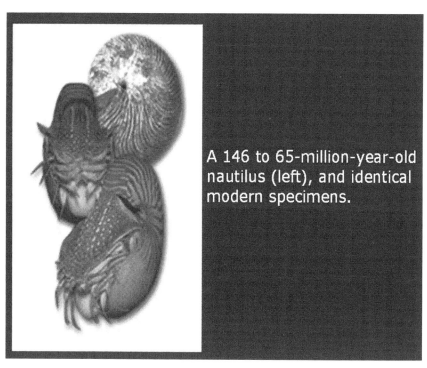

A 146 to 65-million-year-old nautilus (left), and identical modern specimens.

Millions of years for fossils to form? How does that idea square with this ichthyosaurs, fossilized while giving birth to its baby which is still partially in the birth canal.

This is what a Common house fly looked like 54 million years ago !!!

A fish fossilized while trying to eat another fish. From the Green River formation.

Here's what a walking stick looked like 54 million years ago.

Here's what a walking stick looked like last year!

The oldest known fossil scorpion, found in East Kirkton in Scotland. This species, known as *Pulmonoscorpis kirktoniensis*, is 320 million years old, yet no different from today's scorpions.

The fossilized insects were trapped in amber, a fossilized tree sap. There are literally tens of thousands of fossilized insects found in amber. As you can see, they are virtually the same as today's insects. There is absolutely no evidence of any kind of evolutionary change in any plant or animal found in the fossil record!

This 170-million-year-old fossil shrimp from the Jurassic period is no different from living shrimps.

Evergreen Live Oak leaf from the Green River formation

Evergreen Live Oak from California

where are the predecessors for all these plants, insects and animals? Uh, the "missing link(s)" are, ah, still missing!

A fossil maple leaf dating back millions of years, and modern maple leaves.

How about the progression from simple to complex? Is that in any kind of order? If we look where it all supposedly started, that is, the Cambrian formation, we find these fossils: *"The early Cambrian fossils included porifera, coelenterates, brachiopods, mollusca, echinoids, and arthropods. Their high degree of organization clearly indicates that a long period of evolution preceded their appearance in the record. However, when we turn to examine the pre-Cambrian rocks for forerunners of these early Cambrian fossils, they are nowhere to be found."*[9] (Dr. Axelrod is a Professor at UCLA)

In other words, the record clearly demonstrates what creationists have been saying all along. Life commenced suddenly, in a moment of time. It didn't need to evolve, it was already fully formed: *"All major categories of marine invertebrates appear in the fossil record during the Cambrian period in geologic history. The second point to be noted is that all these marine invertebrates appear in a fully formed condition. Further, they are essentially unchanged to the present time."*[10]

Evolutionist Richard Dawkins puts it this way: *"It is as though they were just planted there, without any evolutionary history."* While George Gaylord Simpson said of the absence of the pre-Cambrian fossils: *"It's a major mystery of the history of life."*[11]

Along these same lines another famous person, Napoleon Bonaparte, said: *"The nature of Christ's existence is mysterious, I admit; but this mystery meets the wants of man - reject it and the world is an inexplicable riddle; believe it, and the history of our race is satisfactorily explained."*[12]

"Scientists have let prejudice interfere with their scientific judgment. The unfortunate result was that two generations of students were taught that science had fossil evidence proving that man had descended from apes."[13]

Philip Johnson, a Berkley law professor specializing in the logic of arguments, subjects the scientific support for evolution to careful scrutiny. Mr. Johnson spent years researching the scientific evidence for evolution, and writes of his findings: *"In short, if evolution means the gradual change of one kind of organism into another kind, the outstanding characteristic of the fossil record is the absence of evidence for evolution."*[14]

Another noted paleontologist, Austin Clark, one time curator of paleontology at the Smithsonian Institution, writes of the fossil record of different animal groups and what it reveals: *"Thus so far as concerns the major groups of animals, the creationists seem to have the better argument."*[15]

You have just read the testimony of some evolutionists, and as you can clearly see, the fossil record does not uphold in any manner of a gradual transitional advancement of life on this planet. A transitional advancement must be there if evolution could have any merit in a world of true science. What the fossil record does show is that there was a sudden and massive extinction of almost all life forms on the planet. And that these life forms were instantly buried in tons of water-laid sediments. Further, that fossils are found worldwide is strong evidence for a global deluge in which all life on the earth perished except those animals which were safely on board the ark of Noah. Again, I quote Philip Johnson and his conclusion of what the fossil record clearly demonstrates: *"The general picture of animal history is thus a burst of general body plans followed by extinction."*[16]

The last bit of evidence I want to cover before going on to the next chapter is also a fossil. This fossil is known as a polystrate fossil. I want to dedicate this fossil to every atheist and/or humanist teacher or college professor who may have had a hand in destroying the faith of a new Christian student or a young child's faith in the God of creation. These fossils, perhaps more than any other fossil type, stand in the face of evolution and reveal the impossibility of their position. These fossils declare the written Word: ***"In their wisdom they became fools."*** (Romans 1:22)

Polystrate Fossils

Several years ago fossilized trees were found protruding through 30 to 50 ft. of the geologic column. The fact that the trees penetrate several hundreds of millions of years of geologic time presents a serious problem for the uniformitarian time table. These fossil trees are found in coal seams and are called "kettles." Geologist, John Morris writes: *"I have personally been in many underground coal mines and with one exception, saw polystrate trees or kettles in each of them. Dramatic examples are sometimes found in areas where the coal cross-section is exposed by erosion or by open pit-mining."*[17]

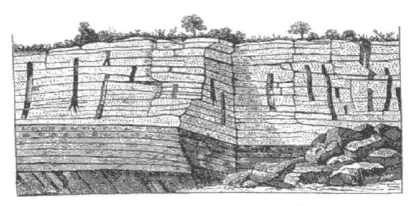

fig. 442. — Vertical trees in the Coal Measures sandstone, St.-Étienne, France. From Credner's *Elemente der Geologie*.

In several instances the trees had been growing on the site and the sediments laid down around the tree as it stood in place. These sediments (making up differing rock layers) were all laid down in a short period of time by a world wide deluge. They are layered because of the different densities of the material which make up each layer of rock stratum. The heavier the material, the faster it would fall through the water, with the lightest coming down last. In this way, layer upon layer of sediment was laid down, ultimately to be pressed into rock by the pressure of water standing above it.

"It is clear that trees in position of growth are far from being rare in Lancashire (Teichmuller, 1956 reaches the same conclusion for similar trees in the Rhein-Westflen Coal Measures), and presumably in all cases there must have been a rapid rate of sedimentation."[18]

Another kind of polystrate fossil appearing on the scene are schools of small fish which have been fossilized together but are in different layers of stratum. Again if as evolutionists assume, these layers are millions of years apart in their respective formations then these small fish of the same species would have had to come along millions of years apart and suddenly be buried in the same spot, only in the next layer up. Sounds kind of fishy doesn't it? It's far easier to believe they were buried at the same time and came to rest in differing levels of the sediments. This explanation also requires far less faith to believe than the one presented by evolutionists.

In closing this chapter on the fossil record and what can be learned from it if we aren't to biased, I quote several of the world's prominent scientists and what they say the fossil record tells us.

Professor Louis Agassiz of Harvard, *"America's leading biologist"* according to Steven J. Gould, wrote the following:

"Species appear suddenly and disappear suddenly in progressive strata. That is the fact proclaimed by Palaeontology."[18]

"Increasing diversity and multiple transitions seem to reflect a determined and inexorable progression toward higher things. But the paleontological record supports no such interpretation. There has been no steady progress in the higher development of organic design. We have had, instead, vast stretches of little or no change and one evolutionary burst that created the entire universe."[19]

"New species almost always appeared suddenly in the fossil record with no intermediate links to ancestors in older rocks of the same region."[20]

"In any local area, a species does not arise gradually by the steady transformation of its ancestors; it appears all at once and fully formed."[21]

"We are forced to the conclusion that most of the really novel taxa that appear suddenly in the fossil record did in fact originate suddenly."[22]

"Modern gorillas, Orangutans, and chimpanzees spring out of nowhere, as it were. They are here today; they have no yesterday."[23]

Reference

1. F.H. Rhodes, "Fossils" p.10
2. Pierre Grasse, "Evolution of Living Organisms" p.4
3. C. O. Dunbar, "Historical Geology' (2ND edition), p.47
4. D. Gish, "Evolution The Fossils Say No" p.31

5. C.O. Dunbar, Ibd., p.47
6. S. Stanley, "Macroevolution, Pattern and Process" p.39
7. S. J. Gould, "The Ediacaran Experiment" Natural History, Vol. 39, p.23
8. D.S. Woodruff, "Science" Vol. 128, p.7
9. D.I. Axelrod, "Science" Vol. 128, p.7
10. J. Wiester, "The Genesis Connection" p.161
11. G. Simpson, "The Meaning of Evolution" p.18
12. Encyclopedia of Religious Quotes
13. J. Wiester, Ibd., p.130.
14. P. Johnson, "Darwin On Trial" p.50
15. Clark, "Animal Evolution" Rev. Biology, p.523
16. P. Johnson, Ibd., p.55
17. John D. Morris, "The Young Earth" p.101 (1994)
18. Prof. Agassiz, "The American Journal of Science, p.142 (1860)
19. S. Loria, S. Gould & Singer, "A View of Life" (1981) College Textbook
20. Section 3.1(a), "A View of Life"
21. S.J. Gould, "Evolution's Erratic Pace"
22. Ayala & J. Valentine, "Evolving: The Theory and Processes of Organic Evolution" p.266-267
23. D. Johansen & M. Edey, "Lucy: The Beginnings of Humankind" p.363

Notes

ALLY OOP

No writing concerning the theory of evolution would be complete without a study of the "cave man." Every American from a small child to adult knows about the cave men, those early forerunners of modern man, the hunched over, half man, half ape, knuckle dragging kind. He had animal skin clothing, a big "ole" club and he drug his girlfriend around by the hair.

Have you ever noticed how evolutionists in their books and on public television programs, while depicting early man, always try and make him look as ape-like as possible, while at the same time, make apes look as human-like as possible? It's an excellent tool to further instill the deception of evolution into the minds of an unsuspecting public.

As we discuss the evolutionary chain leading to the final stages of man, the first place to start is the insectivores. Insectivores were small mammals which ate insects. After millions of years, insectivores (it is believed) evolved into primates. The primate family consists of two suborders, the prosimi and the simii. It is in the order of simii (anthropoidea) we find monkeys, apes and man. These are in turn divided by three super-families: the ceboidea, or new world monkeys, the cercopith ecoidea or old world monkeys, and the hominoidea, or apes of which man also is included.

The first "caveman" I want to discuss is Hesperopithecus. As a high school student, I still remember learning about Hesperopithecus. This fine specimen, it was taught, lived more than a million years ago on the high plains of Nebraska, hence the common name "Nebraska man." Nebraska man was discovered by a man whose name was Harold Cook. Mr. Cook, while doing some digging found a tooth. This single tooth ultimately found its way before the greatest anthropologists of the day and was declared by them as proof positive of the existence of pre-historic man in America.

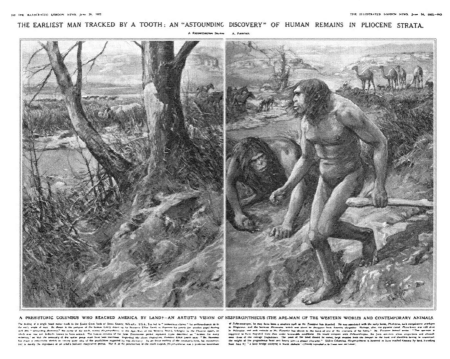

This is the picture that appeared in the London News in 1922 of what they believed Nebraska man looked like. The entire phony story, as you read earlier was based on the finding of a single tooth.

The story of Nebraska man was also promoted by Sir Grafton Elliot Smith, Professor of Anatomy at Manchester, who drew the illustrations of not only Nebraska man, but also his wife. These pictures appeared in the Illustrated London News and were hailed with great pomp as a victory for modern science.

Several years later another tooth just like the first one was found on the Nebraska plains. In this find the entire skeleton was also discovered and it turned out that Nebraska man wasn't a man at all, in fact, he wasn't even an ape! No, the tooth of Hesperopithecus belonged to an extinct pig. However, as is usually the case, little publicity was given to the error.

Piltdown man: (Eoanthropus Dawsoni) also known as the "Dawn man," was discovered in 1911 by Charles Dawson. The discovery of Piltdown won world acclaim. This particular skull was responsible for damaging the credibility of the creationists (during the famous Scopes trial) side by "proving" what seemed to be clear support for the evolution theory. *"In 1953 a fluorine test was conducted on the skull of Piltdown and it was revealed to be a hoax. The jaw came from a modern ape and the human skull (top part) was much younger than the gravel in which it had been found. The jaw had been stained to make it look old and the teeth filed to make them look human."*[1]

Peking man: (Sinanthropus Pekinensis), discovered in 1918 by J. Gunnar Anderson at Chikusan, China. The original discovery was also a single tooth. Later on, ten skeletons were supposedly found in a cave. I say "supposedly" because the bones of Peking man disappeared somewhere between 1941 and 1945 and have never been seen since. In his book "Apemen" Martin Bowden refers to the missing bones as a *"disappearing act,"* while Duane Gish doubts that the cave where the bones were found even exists. So today, a so-called ape-man exists in the high school textbooks that only its discoverers have seen. This is proof for evolution?

Java man: (Pithecanthropus Alalus), discovered in 1887 by Eugene Dubois, a Dutch physician on the island of Java. Before his death and after he had convinced most evolutionists as to the man-like affinity of Pithecanthropus, Dubois himself changed his mind and declared that his Java man was nothing more than a large gibbon. It has since been shown that the skull is indeed that of a large monkey.

"The Naked Ape" (Ramapithecus), was promoted through the pages of National Geographic Magazine. The only fossils ever found of this so-called ape-man are upper and lower jaws and some teeth. Richard Leakey said of Ramapithecus, *"....inescapably it is a matter of faith and this makes the whole problem more challenging and more exciting."*[2]

When faith is required to believe in an idea, then it is moved from the realm of science into that of religion. Remember, it is the evolutionists who continually refer to creation science as being "religion," but as you shall see before finishing this book, evolution, in reality, is nothing more than a religious ideology.

Neanderthal man: (Homo Neander-thalensis), the bones of Neanderthal were first discovered in a cave in the Neanderthal valley near Dusseldorf, Germany. He is portrayed as a semi-erect brutish sub-human. This misconception of Neanderthal is most likely due to the bias of evolution minded paleoanthropologists.

Since the original find, new bones of Neanderthal have been found in caves in Germany and Iran. These bones indicated that Neanderthal didn't walk stooped over as first suspected, but rather walked upright like modern man. Upon further investigation it has been found that the skull bones and other skeletal parts were deformed by a type of arthritis caused by the damp living conditions found in caves. It is also known that these cave dwellers suffered severely from rickets, caused by a deficiency of vitamin D.

It is now known that Neanderthal man was fully erect and in most cases was indistinguishable from modern man, his cranial capacity even exceeded that of modern man. It is said that if he were dressed in a business suit, and were to walk down one of our city streets he would be given no more attention than any other individual. Today he is classified Homo Sapien, fully human.

Cro-Magnon: discovered in 1868 in France. Evolutionists today don't like to bring up the subject of Cro-Magnon (Kro-man-yon). First, he had artistic ability, for he painted pictures on the walls of his cave home. This artistic talent exceeds that of many people alive today. The Book of Knowledge says of Cro-Magnon: *'Fossils related to the Cro-Magnon are the oldest of which we can say positively that they are not only of men but of men of the same species as ourselves."*[3]

Lucy: (Australopithecus Africanus), discovered by R.A. Dart in Africa. Australopithecus had been thought of as a bipedal (walking upright) primate. In his book titled *"The Dragons of Eden"* author Carl Sagan has this creature pictured as walking upright. It can now be shown that such is not the case. Australopithecus was an ape and contrary to Sagan's book, this ape didn't walk upright.

Lord Solly Zuckerman, a famous British anatomist, has studied all known Australopithecine fossils that are important for the last 15 years. In fact, probably no other person has studied these forms more than Zuckerman. His conclusions were that Australopithicus was an ape. Furthermore, Lord Zuckerman states that Australopithecus was not a descendent of man.

The public broadcasting system recently (March 1994) completed a series on its Nova program, *"In Search of Human Origins."* Mr. Johanson steadfastly maintains that Australopithecus (Lucy) is our ancestor and made this statement on the program: *'We should embrace this incredible similarity we have with the apes. We should have a tremendous respect for our closest living biological relative."*

Most anthropologists today however reject Lucy as being related to modern man. C. Oxnard said of Lucy, *"Australiopithecus could not have been ancestral to Homo, they were more proficient in the trees and more different from modern Homo in their form of bipedalism than was previously believed."*

If that's not enough of a problem, consider the latest fossil find at Olduvai Gorge in Africa by DeVore, who just found the fossil remains of Homo erectus in the same rock stratum as Australopithecus! You might want to ask yourself this question: what is modern man doing in the same rock layer as the ape he evolved from? Richard Leakey said of this find: *"Either we toss out this skull or we toss out our theories of modern man."*

The Nutcracker man: (Zinjanthropus Bosi), discovered in 1959 in the Olduvai Gorge in Tanzania by Louis and Mary Leakey. The Leakey's research was sponsored by National Geographic. A combination of rather extravagant claims by Leakey for his find combined with publicity through the pages of National Geographic magazine succeeded in conveying the idea that Leakey had made a unique and momentous discovery at Olduvai.

Today however, even Leakey admits that his "Nutcracker man" is nothing more than a variety of Australopithecus. Another ape!

One of the leading anthropologists in the world said concerning man's relationship to apes: *"If you brought in a smart scientist from another discipline and showed him the meager evidence we've got, he'd surely say, 'forget it, there isn't enough to go on."* [4]

Dr. H. Rimmer wrote: *The evidences of an extended antiquity for man are purely hypothetical, and entirely out of the imagination and desire of the sponsor of such evidence."* [5]

Similarly, T. White, a Berkeley anthropologist observes: *"The problem with a lot of anthropologists is that they want so much to find a hominid that any scrap of bone becomes a hominid bone."*

*P*erhaps Paul Shipman summarizes the quandary evolutionists have gotten themselves into the best: *"....where is the ancestral hominid species? The best answer we can give right now is that we no longer have a very clear idea of who gave rise to whom: we only know who didn't. This uncomfortable state of affairs can be summarized in three simple statements: (1) Robustus didn't evolve into boisei. (2) Africanus didn't evolve into boisei. (3) Boisei didn't evolve into either africanus or robustus. In fact, we don't even know what sort of 'ancestral species' we're looking for...like an earthquake, the new skull has reduced our nicely organized constructs to a rubble of awkward, sharp-edged new hypotheses. It's a sure sign of scientific progress."*[6]

Finally, I would like to say I think it is quite clear that the only real place cave-men ever existed was in National Geographic magazine who had the audacity to refer to a creationist organization as *"The cracked earth society."* Apparently, those at National Geographic believe themselves to be the sole possessors of real knowledge.

References

1. World Book Encyclopedia, (Vol. 15, p.417), 1973
2. Richard Leakey & R, Lewin, "Origins" p.84
3. The Book Of Knowledge, (Vol. 2), p.646
4. Richard Leakey, "The Making of Mankind"
5. H. Rimmer, "The Theory of Evolution and the Facts of Science" p.18
6. P. Shipman, "Baffling Limb on the Family Tree" p.87-93

Notes

SURVIVAL OF THE FITTEST???

The fittest have been identified by evolutionists as those life forms which have been successful in reproducing after their own kind: *"That all living things vary is indisputable. Those individuals endowed with the most favorable variations, would have the best chance of surviving and passing their favorable characteristics on to their progeny. This differential survival, or 'survival of the fittest,' was termed natural selection."*[1]

I mentioned the peppered moth and how it is widely touted today as an example of evolution in progress. The darker moths survived because they were darker than the lighter variation. Thus they were able to fit the criterion as the most successful of these moths to pass on their "favorable characteristics," so the darker moth became the most dominant. They survived because they were dark in color, and they were dark in color because they survived!

As I mentioned before, natural selection and survival of the fittest in reality mean very little. Sir Karl Popper wrote: *"To say that a species now living is adapted to its environment is, in fact, almost tautological. Indeed, we use the terms 'adaptation' and 'selection' in such a way that we can say that, if a species were not adapted it would have been eliminated by natural selection. Similarly, if a species has been eliminated it must have been ill adapted to the conditions."* (Taken From "The Philosophy of Karl Popper" p.137)

In a similar vein, R. Brady concurs: *"If we substitute 'adaptive complexity' for 'fitness' we are indeed in danger of tautology, since the term 'adaptive,' by presuming some sort of positive adaptation to fit the environment, already means 'fit', and thus it has no explanatory value here."*[2]

"Although natural selection theory fails to explain the origin of evolutionary novelties, its greatest shortcoming in terms of evolutionary theory is that it fails to explain evolutionary diversity."[3]

Even if natural selection made any sense at all, it still would be of no benefit to evolution because natural selection can only chose from pre-existing genes, it cannot create new ones. Thus, if the only peppered moths in existence today were the dark colored ones, what would we have? The exact same thing we would have had several hundred years ago when the darker varieties became dominant; peppered moths then, peppered moths now!

Another example evolutionists continually point out as being a good case for evolution are insects that are born immune to some types of insecticides. A grasshopper, for example, is sprayed with a certain kind of insecticide. The grasshopper doesn't die from the spray because it didn't get enough to kill it but enough to make it sick. Its biological system builds a partial immunity to the spray. It passes this immunity trait on to several of its progeny which in turn are sprayed with just enough insecticide the next summer to make them sick. They recover even more immune than the parent grasshopper. In this way, if the farmer continues to use the exact same chemical insecticide every year, it will become less potent to the intended victim, eventually becoming more or less harmless to them.

I have heard this argument used many times by evolutionists. The argument is quite easy to quell simply by asking the question: "is this changing the species?" The answer of course will be no, this wouldn't change the grasshopper species any more than it does for humans who become immune to different diseases and/or exposure to differing biological chemicals such as penicillin.

Some have argued that survival of the fittest is a description of the strongest, most ferocious of a given species. After all, how could they continue to produce progeny if they weren't strong enough to physically survive? The strongest then would be those animal types which could easily defend themselves or escape their enemies. These most often would be the animals at the top of the food chain: lions, tigers, wolves, and other great predatory animals. The question then is: did the strongest survive?

Consider for a moment the modern elephant which is found in both Africa and India. We know from the fossil record that our modern day elephants are dwarfed by the fossil Imperial elephant which in turn was smaller than the hairy mammoth or the woolly mastodons. Ostriches have been found in fossil form; their size is comparable to the modern giraffe. Fossil pigs have been found 18 feet tall and bears, some 20 feet tall. Some of these made the Alaskan Kodiak appear as a dwarf. These animals were all larger, they were all stronger...but they didn't survive!

The same principle holds true for birds, insects and reptiles. Birds with 20 and 30 ft. wingspans are known in the fossil record. Dragonflies with wing spans of 20 and 30 inches are known. Reptiles 75 and 85 feet long and weighing 40 and 50 tons have appeared in the fossil record...but they didn't survive!

Dr. Colin Patterson, senior paleontologist at the British Museum of Natural History, in a talk he gave at the American Museum of Natural History, Nov. 5, 1981 said that he now realizes that in accepting evolution he had moved from science into faith. In a recent BBC program, Dr. Patterson stated: *".... all we really have of the evolutionary phylogenetic tree are the tips of the branches. All else, the filling in of the trunk and of the branches is simply story telling of one kind or another."*

Patterson is also quoted by Leith: *"There's no doubt at all that natural selection works - it's been repeatedly demonstrated by experiment. But the question of whether it produces new species is quite another matter. No one has ever produced a new species by means of natural selection, no one has ever got near it, and most of the current argument in neo-Darwinism is about this question."*[4]

Similarly Dobzhansky could find not one single shred of datum for the role of natural selection: *"The role assigned to natural selection in establishing adaptation, while speciously probable, is based on not one single sure datum."*[5]

Dr. Henry Morris, president of the Institute For Creation Research said of evolution: *"One of the most amazing phenomena in the history of education is that a speculative philosophy based on no true scientific evidence could have been universally adopted and taught as scientific fact in all the public schools."*

Robert Ardry, an evolutionist who has written extensively on the subject, said of the academic world of evolution: *"The three sciences central to human understanding - Psychology, Anthropology, and Sociology - successfully and continually lie to themselves, lie to each other, lie to their students, and lie to the public at large."*[6]

In summation, I would again like to quote Pierre Grasse, who is perhaps the most distinguished of all French Zoologists: *"Through use and abuse of hidden postulates, of bold, often ill-founded extrapolations, a pseudoscience has been created. It is taking root in the very heart of biology and is leading astray many biochemists and biologists, who sincerely believe that the accuracy of fundamental concepts has been demonstrated, which is not the case."*[7]

References

1. E.P. Volpe, "Understanding Evolution" (5th edition), p.20
2. R. Brady, "Systematic Zoology" p.600
3. Rosen, "Darwin's Demon" p.27
4. Leith, "Are the Reports of Darwin's Death Exaggerated?" p.166 (1981)
5. P. Grasse, "The Evolution of Living Organisms" p.170 (1977)
6. R. Ardry, "The Social Contrast" p.12
7. P. Grasse, "The Evolution of Living Organisms" (1977)

Notes

EVOLUTION THE RELIGION

If the public school system today tried to push a religion onto its students, the parents would rise up in righteous indignation. But if that religion is foisted upon our society in the name of science, well, that's a different story. I am sorry to have to inform you if your children attend the public school system, they are being indoctrinated with the religion of humanism. Evolution is the foundation upon which this religion sits, as well as eastern mysticism, and the so-called "New Age" movement.

In 1963, the Supreme Court declared humanism to be a religion. So why do we allow it in our schools? As just mentioned, we have taken their religious ideology and dressed it up and passed it off as science. Consider this statement made by J. Dunphy in the "Humanist." It explains exactly how the public school system is influencing our children with their religious beliefs. *"I am convinced that the battle for humankind's future must be waged and won in the public school classroom by teachers who correctly perceive their role as the proselytizers of a new faith: a religion of humanity that recognizes and respects the spark of what theologians call divinity in every human being. These teachers must embody the same selfless dedication as the most rabid fundamentalist preachers, for they will be ministers of another sort, utilizing a classroom instead of a pulpit to convey humanist values in whatever subject they teach, regardless of the educational level - preschool, day-care, or large state universities. The classroom must and will become an arena of conflict between the old and the new - the rotting corpse of Christianity, together with all its adjacent evils and misery, and the new faith of humanism."*

Julian Huxley wrote: *"A religion is essentially an attitude to the world as a whole. Thus evolution, for example, may prove as powerful a principle to coordinate man's beliefs and hopes as God was in the past."*[1]

Professor Louis T. More, one of the more vocal evolutionists said: *"The more one studies paleontology, the more certain one becomes that evolution is based on faith alone."*

If that thing which we believe requires faith in order to believe in it, then it becomes a religion, and not science. Evolution carried out to its end has man evolving to a state where he will become god. The very reason Eve ate the forbidden fruit in the garden was to become like God: **"For God knows that in the day you eat thereof, then your eyes shall be opened, and you shall be like God..."** (Genesis 3:5)

Jeremy Rifkin, a Professor at Stanford University wrote these words in his book titled *Algeny: "We no longer feel ourselves to be guests in someone else's home and therefore obliged to make our behavior conform with a set of pre-existing rules. It is our creation now. We create the world, and because we do, we no longer feel beholden to outside forces. We no longer have to justify our behavior, for we are now the architects of the universe. We are responsible to nothing outside ourselves, for we are the kingdom, the power, and the glory forever and ever."*[2]

One of the most outspoken proponents of evolution in our time has been J. Huxely. Mr. Huxely boldly laid out the goals for humanistic evolution in the publication of the "Humanist" when he wrote: *"The unifying of traditions into a single common pool of experience, awareness and purpose is the necessary prerequisite for further major progress in human evolution. Accordingly, although political unification in some sort of world government will be required for the definitive attainment of this state, unification in the things of the mind is not necessary also, but it can pave the way for other types of unification."* And...."*Thus the general philosophy of UNESCO should, it seems, be a scientific world humanism, global in extent and evolutionary in background."*[3]

Make no mistake, atheistic humanism has inundated our entire school system and to some degree even the churches of our land. Proponents of global education are intrinsically tied into New Age religion and evolution. They have disguised their program of indoctrination with names like: "Multicultural Education, International Curriculum Development, Cultural Awareness, Project 2000, and, Welcome to Planet Earth." Richard Sutphen, a leading New Age thinker, summarizes these so-called New Age doctrines as: *(1) the external world and consciousness are one and the same; (2) we are all part of god, so we are god; (3) life is for evolutionary purposes; (4) awareness of one's true self within leads to mastery of one's own reality."*[4]

The ethnic religions of the East, (Hinduism, Taoism, Buddhism, etc.) continuing the polytheistic pantheism of the ancient pagan religions, have long espoused evolutionary views of the universe and its living things, and so merge naturally and easily into the evolutionary framework of New Age philosophy.

Some educators and many leading scientists call for a paradigm shift in which events are viewed holistically and in which scientific laws, constructs, and views of how the human mind thinks and functions are redefined. Many of these scientists and educators strongly advocate the so-called "Gaia Hypothesis" (Gaia was a Greek goddess of the earth - "Mother Earth"). This mystical Gaia Hypothesis assumes that the earth is a living entity which actually induced its own creation through the process of evolution.

It is within the confines of this all encompassing world religious system that they have fit the idea of environmentalism. The primary role of environmentalism is to convince the world's populace that the world's environment is about to be destroyed unless we take drastic measures to save it. Those measures of course will be to persuade the world's populace to come under the leadership of a one world government.

Many educators today, will not only inhibit the teaching of creation science in the schools, they would go further as pointed out by two Iowa professors: *"....as a matter-of-fact, creationism should be discriminated against... No advocate of such propaganda should be trusted to teach science classes or administer science programs anywhere or under any circumstances. Moreover, if any are now doing so, they should be dismissed."*[5]

In closing this chapter, I should point out that not all evolutionists support this kind of restrictive censorship. Provine, a prominent historian of science at Cornell and the co-editor of *"The Evolutionary Synthesis"* writes: *"First of all I said that creationism should be taught along with evolutionism in grade schools and high schools. My motivation for wishing them to be discussed in this way is twofold. First of all I believe strongly in an open discussion of ideas. I do not believe that natural scientists should suppress the creationist point of view and keep it out of the science classroom when creationism is a viable, understandable and plausible theory for the creation point."*[6]

T. Black, a former chancellor of the New York State Board of Regents, agrees with Provine: *"Monopoly in education can become thought control, denying the individual teacher's and student's right to read and listen and think and debate and make decisions and speak out. The views of those who hold the monopoly are exalted, while nonconforming views are suppressed or ignored. Thus education ceases to be education and becomes indoctrination."*[7]

"No teacher should be dismayed at efforts to present creation as an alternative to evolution in biology courses; indeed, at this moment creation is the only alternative to evolution. Not only is this worth mentioning, but a comparison of the two alternatives can be an excellent exercise in logic and reason."[8]

References

1. Julian Huxley & J. Bronowski, "Growth of Ideas" p.99
2. J. Rifkin, "Algeny" p.195
3. J. Huxley, "A New World Vision" (The Humanist, Vol. 39, April 1979)
4. E. Buehrer, "The New Age Masquerade: The Hidden Agenda in Your Child's Classroom"
5. Scientists and Educators Discriminate Unfairly against Creationists?" Journal of the National Center for Science Education, p.19
6. Provine, "Scientists Abandon Evolution" Contrast, March - April 1982
7. T. Black, "Straight Talk about American Education" p.40
8. P. Davis & Solomon, "The World of Biology" p.414

Notes

AND THERE WAS LIGHT

In 1917 one of the world's great physicists developed an equation that was to astound the nations. The equation $E=mc^2$ seemed simple enough outwardly, but locked within its mathematical corridors were awesome secrets which even today aren't fully realized.

Einstein's theory of relativity is an assertion of the equivalence of mass and energy. It was this equation that opened the doors to the world of atomic energy and a few short years later "energy equals mass times the speed of light squared," was realized as the first atom bomb was detonated in a secret test area at White Sands Proving Grounds in New Mexico. Man held within his hands for the first time in history the ability to destroy all flesh on the face of the globe. This new ability gave credence for the first time to a chilling end time's prophecy given by Jesus Christ. While talking to His disciples about the conditions that would be existent on the earth just prior to His return, He said: *"....and except those days be shortened, no flesh would be saved."* In other words, mankind would be on the verge of annihilating himself through global war, except that Christ Himself would cut the time off and return to save man from total extinction.

When Albert Einstein developed his famous equation he was an atheist. Shortly afterward a Russian mathematician, Alexander Friedmann, discovered an error in the master's mathematics and made the correction. The error that Mr. Friedmann found caused Einstein to overlook additional solutions to the equation. The additional information to the equation had profound implications for it clearly showed the relationship of mass to light and time. These elements are the products of creation, and a creation, of course, necessitates a creator.

The indication astonished the scientific community. Not only did the universe have a beginning, but time and light played an integral part in that beginning, but who was the Creator? Dr. Einstein became despondent, and in a letter to colleague Willem de Sitter he wrote: *"To admit such possibilities seems senseless."* Clearly Einstein was troubled by the idea of a higher being and even hoped that somehow he would be proven wrong.

In 1919 his equation was shown to be correct when it was proven that light is bent by gravitational fields. Dr. Einstein finally conceded, the universe wasn't the result of chance after all, but rather a definite order had been established by an intelligent being and all of His creation was in subjection to the laws He established which govern the universe.

When we turn to the first page in the Bible we realize the scientific accuracy then of the first three words: **"*In the beginning.*"** There was a beginning, there was a creation of time, and this truth is born out elsewhere in the Bible as we find in II Timothy 1:9 these words: **"*Who has saved us and called us with a holy calling not according to our works, but according to His own purpose and grace which was given to us in Christ Jesus before time began.*"** And, **"*....in hope of eternal life which God, Who cannot lie, promised before time began.*"** (Titus 1:2) Indeed, God had a plan for mankind before He ever created the world or time!

The theory of relativity clearly demonstrates that time is a created entity, and Scripture upholds the science, but how does light fit into the equation? Let's begin with the first verse in the book of Genesis: **"*In the beginning God created the heavens and the earth. The earth was without form and void; and darkness was upon the face of the deep. And the Spirit of God was hovering over the face of the deep. Then God said, 'Let there be light, and there was light.*"** (Gen. 1:1-3)

You have just read about the creation of time, light and mass (the earth), all three ingredients to Einstein's famous equation! At the time Einstein developed this equation and the world marveled at his genius, the ancient Hebrew Scriptures we call the Old Testament had, four thousand years before, already spelled it out! This is no quirk, no mistake, or a coincidental happenstance, there is an intentional interjection of the word *"light"* as a definite entity to the physical creation! You see, the sun had not yet been called into existence by God, as a matter-of-fact, the sun wasn't created until the fourth day (Genesis 1:14), thus the word *"light"* portrays an integral part of the physics of God's creation of the universe.

Some have questioned whether this idea is born out elsewhere in the Scripture. The answer to that is a resounding yes. Let's read on: **"The day is Yours** (God's) **and the night is Yours; You have prepared the light and the sun."** (Psalm 74:16) And: **"I form the light and create darkness..."** (Isaiah 45:7)

Notice the two verses just quoted both indicate that light is a specific entity and not necessarily the natural result of a physical happening such as the sun shining. Today, of course, we can observe natural light being produced by a physical happening such as the light of a candle. There are also various kinds of artificial lights produced by differing electrical currents. There is also light energy that is invisible to the naked eye; x-rays, and radio waves being two examples of invisible light waves. But all light, whether artificial or natural, visible or invisible, produces magnetic energy of differing wave lengths. Light energy is ultimately the glue that holds the universe together. Outside of God, if there is no mass, there can be no light, and by the same token, if there is no light, there can be no mass. Thus, the evolutionist, who must try and explain the beginning of the cosmos (energy) from nothing, must of necessity maintain a high degree of faith in his belief, for there is absolutely no scientific support for him to stand on.

As already mentioned, Einstein was troubled by the time, light, mass equation, as it clearly demonstrated a moment of genesis for the cosmos, and precludes irrevocably the idea that the cosmos might have existed forever. Today we know through solar observation that the sun is shrinking at the rate of about 2 feet per hour which is equivalent to losing .05% of its diameter every 100 years.[1] Quite obviously then, the sun couldn't possibly have been in existence very long, otherwise it would have burned itself out long ago.

The sun is shrinking because it is converting part of its mass into energy. Because of this conversion, we should be able to rewrite Einstein's famous equation to show [energy produced] = [mass eliminated] X [speed of light^2.] Thus we could say that the sun is losing some of its mass through energy conversion, while the earth and other planets of our solar system are gaining a portion of what the sun is losing.

The sun's light is energy therefore it is equal to mass. While we may not think of light as being equal with mass, consider for a moment that the sun's light exerts pressure on anything that obstructs it. If it were possible to gather one square mile of sunlight, we would find that it would have a weight of about three pounds, while all the sunlight falling on the earth's surface would weigh about 87,700 tons.[2]

So, as the sun shrinks, losing 2 feet of surface mass each hour, the earth and other planets in our galaxy will increase in size as they gain what the sun loses. In fact, if we were to extrapolate backward in time, and add two feet of surface mass for each hour that has gone by, and do so for 25,000 years, it can be shown that the sun would have been so large it would have boiled away all the oceans of the world. Continuing on backward in time to a mere one million years, the sun would have been so large it would have heated the earth to a molten ball. Still continuing backward in time to two hundred ten million years, the sun would have been so large its surface would have engulfed the earth in its sphere!

It should be pointed out that evolutionists extrapolating back to the "Big Bang" are extending 95 times further than our shrinking sun event.

I mentioned earlier that light is bent by gravitational fields. Gravitational fields also have a direct relationship to time (remember, time, light and mass are all interconnected), in fact, time is "warped" by gravity. The stronger the field, the more time is affected. This time distortion is known as "gravitational time dilation."

In his book titled *"Starlight and Time"* author Russell Humphreys, a nuclear physicist at Sandia National Laboratories, writes concerning gravity and time distortion: *"....an atomic clock at the Royal Observatory in Greenwich, England, ticks five microseconds per year slower than an identical clock at the National Bureau of Standards in Boulder, Colorado, both clocks being accurate to about one microsecond per year. The difference is exactly what general relativity predicts for the one-mile difference in altitude.* (ie., increase in mass)

"Which one is showing (or running at) the 'right time'? Both are - in their own frame of reference. There is no longer any way to say which is the 'correct' rate at which time runs - it all depends on where you are in relation

to a gravitational field. A large variety of more precise experiments have confirmed gravitational time dilation to an accuracy of better than one percent, it's for real!"

Dr. Humphreys continues: *"What this new cosmology shows is that gravitational time distortion in the early universe would have meant that while a few days were passing on earth, billions of years would have been available for light to travel to earth."*³

Are you getting the picture? *"....for in six days the Lord made the heavens and the earth, and on the seventh day He rested...."* (Exodus 31:17) The universe was created by the Most High God in six literal 24 hour periods, in fact, so that there could be no mistake, He spells it out in solar time: ***"So the evening and the morning were the first day."*** (Genesis 1:5) This creation period took place approximately seven thousand years ago (in earth years), but billions of years could have elapsed in terrestrial time. Whose clock is correct? Both, in their own reference frames! Keep this thought in mind because I'm going to come back to it in a later chapter.

There are several more intriguing properties of light that I want to get into before moving on to the next chapter. Earlier I stated: "Light energy is ultimately the glue that holds the universe together." Let me explain. Cosmologists today are struggling with the question as to what holds the universe together. Computations have clearly shown that there isn't enough mass in the cosmos to hold it together, in fact, they say, 90% of all the mass needed to hold it together is missing!

Remember Sir Isaac Newton's law of gravity? This law states that every particle in the universe will attract every other particle with a force that varies directly as the product of their masses. In other words, the larger a bodies mass, the more gravitational attraction it will have on a differing body. The question though is...where does gravity come from? The Genesis account of creation clearly pre-dates the theory of relativity by thousands of years, likewise, the answer to what holds the universe together pre-dates any gravitational theory by two thousand years. But before we get to that verse, let's zoom in on one of the tiniest elements of mass known; the atom.

Everything that exists is made of atoms including the air you inhale and the air you exhale, the stars, and all the planets, indeed the entire universe, is made from these tiny spheres of wonder. Atoms are so tiny that billions could easily fit on the head of a pin. I once heard a nuclear physicist lecturing on the size of atoms. The description he gave went something like this.

Imagine a drop of water which just fell from your kitchen faucet into a saucer. You pick the saucer up and set it on the table. You are in possession of a magic pair of tweezers which will allow you to reach into the drop of water and grab one of the atoms making up the drop. You let the atom fall to the floor. As it does, it is increased to the size of a grain of sand. It is your intention to extract every atom making up the drop of water and increase them all to the size of a grain of sand. If this were humanly possible, by the time you finished, you would have enough sand to build a highway 3 feet deep and 15 feet wide, from Los Angeles, California, to the city of New York!

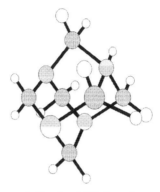

A Typical Molecule

In the early nineteenth century English chemist John Dalton noticed that elements always combine in definite ratios of masses. To explain this observation, Dalton suggested that matter is made up of small particles called atoms. Thus, John Dalton was credited with advancing the idea that all mass is made up of these tiny spheres. But wait a minute! Shouldn't the Apostle Paul have gotten the credit? He wrote: ***"By faith we understand that the worlds were framed by the word of God, so that things which can be seen, are made of things which cannot be seen."*** (Hebrews 11:3)

Clearly Paul was directed by God to write this verse which Dr. H. Rimmer described as: *"....the most concise statement of atomic theory that the human pen could write in one short paragraph."*[4]

The major components of these tiny spheres are: protons, neutrons, and electrons. Protons are positively charged particles, neutrons have no charge and electrons are negatively charged. Electrons are about 2000 times smaller than protons, but contain the same magnitude of charge. Atoms combine together to make up molecules.

Most molecules have two or more atoms. Some molecules have hundreds of atoms. Atoms attach themselves together by the electrons. This orderly and regular behavior of attaching themselves is governed by a number called valence. Different atoms have valence of 0,1,2,3, or 4. There is a strict rule which governs valence, every number must be used, thus an atom with a valence of 0, cannot combine with other atoms which are different because they are inert.

An example of different atoms combining would be one atom of magnesium (valence of 2) combining with 2 atoms of chlorine (valence of 1) forming a magnesium chloride molecule. Another compound that is well known would be the combining of oxygen and hydrogen atoms which, of course, makes water.

This brings us to the crux of our discussion of atoms and light. The spinning of the subatomic particles within the nucleus of an atom generates an electrical field which bonds the atom together. This electrical bonding is called the strong nuclear force, and is literally what holds the atom together. The electrical energy is also what holds the electron in its orbit.

The electron, in turn, combines with other atoms to form molecules. Electromagnetic energy (light), in tiny particles known as "pi mesons," is what causes the spin of the subatomic particles. A particle of light of a definite frequency is called a photon or a light quantum. A photon is the smallest particle of light energy possible. Photons along with pi mesons are produced in a subatomic interaction, where particles of light mass are emitted by slowly decaying heavier particles. Thus when some of the heavier particles called hyperons decay, they emit photons, while others give off pi mesons. Pi mesons are very unstable short-lived particles. This means that from the time a pi meson appears until the time it decays into another particle, it exists for about forty-millionths of a second. Grolier's book of Popular Science says this about this interaction: *"Subatomic particles interact in several different ways. One type of interaction involves the light particles called pi mesons, or pions. It has been found that both protons and neutrons absorb and emit pi mesons rapidly and constantly. This so-called pi meson exchange is the 'cement' that holds protons and neutrons together in the nucleus."*[5]

There is absolutely no place in the universe where one could go to escape light energy. The deepest mines on our planet are continually bombarded with invisible light energy called neutrinos, and probably other forms of light we don't even know about yet. It is theorized that neutrinos might provide the missing mass that is responsible for holding the cosmos together: *"Big bang neutrinos are nearly as numerous as photons of light. They could therefore account for more than 90% of the mass of the universe."*[6]

Cosmologists know that the universe is filled with energy. Copious amounts of light flood into it from an unknown source. Evolutionists, of course, believe this is left over energy from the "big bang." They refer to it as "background radiation."

In the book titled *"The Structure of Space"*, author J. Solomon writes: *"The source of these waves is a total mystery. They have turned out to be far stronger than astronomers could have foreseen, although still barely detectable, far more then a thousand suns would have to collapse into nothingness every year to radiate so vast an amount of this elusive form of energy. All the light and radio waves emitted from our whole galaxy do not equal its volume and the only explanation offered so far seems to touch the borderline of pure fantasy."*[7]

I guess Job shouldn't feel so bad! Here we are four thousand years later and modern science still can't answer the question he was asked. This is the way God put it to Job: **"Have you comprehended the breadth of the earth? Tell Me if you know all this. Where is the dwelling place of light? And darkness, where is its place...."** (Job 38:18,19) Poor Job, he could only stammer and finally he cried out: **"I know that You can do everything, and that no purpose of Yours can be withheld from You. You asked, 'who is this who hides counsel without knowledge?' Therefore I have uttered what I did not understand, things too wonderful for me, which I did not know."** (Job 42:2,3)

Perhaps arrogant man should take a lesson from Job. Man doesn't know everything, indeed, in reality, mankind really knows very little about the cosmos and its origin. If he were standing right beside God in the beginning when He created it, he still wouldn't understand what it was that he saw.

So, the universe is being flooded with vast, unimaginable quantities of light. This light is in reality what holds the whole of creation together. No one knows where it comes from, nor does anyone understand exactly how it works. There is no book, or group of books which tells us that light is what holds everything together....except one book! Let's turn now to its pages because it tells us exactly what we are looking for.

In John 8:12, Jesus declared that He **"is the light of the world."** And the light of God falls upon everyone. (Job 25:3) In I Timothy 6:15, & 16, Paul while talking about the majesty of God wrote: **"....which He will manifest in**

His own time, He who is the blessed and only Potentate, the King of kings and Lord of lords, who alone has immortality, dwelling in unapproachable <u>light</u>, whom no man has seen or can see...." King David, in the book of Psalms writes an exhilarating description of God: *"O Lord my God, You are very great : You are clothed with honor and majesty, Who cover Yourself with <u>light</u> as a garment...."* (Psalm 104:2) David then makes this awesome statement: *"O God of our salvation, You who are the confidence of all the ends of the earth, and of the far-off seas; Who established the mountains by His strength, being clothed with power...."* (Psalm 65:5,6)

What is He clothed with? Light or power? Both you say! Exactly...light is POWER! It is energy of the highest order. Now, turning to Habakkuk 3:4 we read this amazing description of the God of light: *"His brightness was like the light; He had rays flashing from His hand, and there His power was hidden."* The prophet Jeremiah writes: *"He has made the earth by His power...."* (Jeremiah 10:12) He made the earth by His power! What does it say His power is? The prophet Malachi referred to God as *"The <u>Sun</u> of Righteousness."*

In summation: the product of the decaying hyperon produces the pi meson. This exchange *"is the cement that holds protons and neutrons together in the nucleus."*

The electromagnetic force (light energy) on the other hand, possesses an unlimited range. It is virtually everywhere and there is no escaping its energy particles. The exchange particle for the electromagnetic force is the photon. The electromagnetic force affects <u>all</u> charged particles. It is the electromagnetic force that holds the electron in its orbit around the nucleus of the atom. And the electron, in turn, binds all atoms together to form molecules.

This "cementing" together of all subatomic and atomic particles is accomplished through the power of electromagnetic energy of one kind or other. I also pointed out, that the universe is being flooded with light energy so vast and powerful that any *"explanation offered so far seems to touch the borderline of pure fantasy."* Scripture clearly states that God is Light (see John 1:1-5) and by His power, the universe was made. So what holds it together?

"He (Christ Jesus) *is the image of the invisible God, the firstborn over all creation. For by Him all things were created that are in heaven and that are on earth, visible and invisible, whether thrones or dominions or principalities or powers. All things were created through Him and for Him. And He is before all things, and in Him all things <u>consist.</u>"* (Colossians 1:15-17).

The Greek word for *"consist"* (sunistemi) literally means *"to band together."* In other words, through Jesus Christ, all things in the entire creation are bound together. By His power! What is His power?

References

1. Physics Today, "Analyses of Historical Data Suggests Sun is Shrinking" Sept. 1979 p.17-19
2. N. McAleer, "The Cosmic Mind Boggling Book" 1982 p.3
3. Russell Humphreys, "Starlight and Time" 1994 p.11-13
4. Harry Rimmer, "The Harmony of Science & Scripture" p.92
5. Golier Publishing Co. "Book of Popular Science" Vol. 1, 1977 p.3136
6. R. Gore, "The Once and Future Universe" National Geo. Vol. 163 p.744, 745
7. J. Solomon, "The Structure of Space" p.167

Notes

THE CREATION MODEL

The Christian high-schooler squirmed as the skeptical science teacher tried to hold back his laughter. "And just where did this God come from?" he queried. "No where sir," the student replied, "God has always existed." "But" bellowed the teacher, "that statement is not scientifically tenable." "May I ask you a question," asked the student. "You may, but make it quick, this is a science class and we don't have time to talk about religious opinions." "Okay, just one question. Where did the matter come from that makes up the universe?"

The smile began to fade from the teacher's face. "Why, matter has always existed," replied the teacher. "I'm sorry sir, but that is a contradiction of the Second Law of Thermodynamics, it is also contrary to Einstein's Special Theory of Relativity."

And so it goes, each side battling back and forth making claims and counter claims, but which one is right? As you have already read in the first half of this book, there is precious little evidence to support evolution. As a matter of fact, from the mouths of the evolutionists themselves, they rely on faith that someday their wild postulations will be proven true. Let's take a look at the science we call Creation Science, and see what factual evidences we can come up with in support of our brand of science.

The evolutionist must believe that nothing produced the vast array of complex systems we see in operation within our cosmos, or they must believe that it has always existed. In the preceding chapter I briefly discussed the make-up of atoms and what holds them together. But let's zoom one more time back into the atom. This time we'll take a closer look into the subatomic nucleus; here we will find another world, a world of tremendous complexity and beauty, a world of dimension that simply boggles our simple minds. Are these tiny spheres a product of random chance? You be the judge.

The nucleus is made up of approximately 200 different particles. These particles all spin, and each has an antiparticle; that is, for every particle there is another particle of the same mass and spin but possessing an opposite charge. An example of this would be an electron with a negative charge and its counter opposite, the positron with a positive charge. When one particle meets its antiparticle, both disappear and are replaced by energy. This process is known as "pair annihilation." There are some particles which have a rest mass of 0. That is, if they stopped spinning, they wouldn't exist. They have mass only by virtue of their energy (spin), hence our equation $E=mc^2$. Remember, energy equals mass?

Now, ask yourself this question. If there is no God and the universe came into existence from nothing; how did nothing take nothing, and cause it to spin until it became something, and then place it within the nucleus of an atom with more than 200 other particles? If you are an atheist, your science is untenable.

There are times when atoms produce light energy. This is accomplished when an atom absorbs more energy than it can handle. When this happens, the electron is bumped to a higher orbit (that is, it moves further away from the nucleus) where it then dumps the increased energy load by giving off light. A good example of this energy - light exchange would be neon gas. Neon, placed inside a glass tube and charged with electricity, causes the atoms which make up the gas to glow. As a matter-of-fact, light energy itself can charge atoms to a point where they will radiate light. An outside light source falling on an atom will cause the electrons to give off more light because of the increase in energy. The luminous dial on your watch, for example, will glow brighter if it has been exposed to a bright light. It's also a law of physics that any addition of energy to a particle will result in an increase in mass of that particle.

Atoms themselves are a form of light energy, and since everything that exists is made from atoms, all of creation could be made to radiate light. Scripture has a very sophisticated way of saying this same thing of God's handiwork: ***"Every good gift and every perfect gift is from above, and comes down from the Father of lights."*** (James 1:17)

Again, in the book of Romans we read this about God's creation: *"**For since the creation of the world His invisible attributes are clearly seen, being understood by the things that are made....**"* (Romans 1:20) God's *"invisible attributes are clearly seen"*. His entire creation, made from invisible atoms is clearly seen. This is also an indictment against sinful man who ignores the order and complexity of the visible universe, believing it all to be an accident. In fact, the verse just quoted above continues on by saying: *"**....they are without excuse.**"*

The atheist must believe that nothing created all the atoms in the universe, then arranged them in such a way that light would come into existence??? Wait a minute, if there was no light, there could be no atoms, and, if there were no atoms, there could be no light! Thus, "nothing" is left with the irremediable task of arranging chaos into order! This is total foolishness, in fact, the verse just quoted in Romans continues on by saying: *"**Professing to be wise, they became fools.**"* (Romans 1:22)

Consider also for a moment, if God created everything that exists and recorded the very beginning in His Word, there would be no such thing as "prehistoric," for the *"**Father of lights**"* was faithful to record for us the very beginning. He knows how it happened because He was there! Science is only a study of that which He did in the beginning. Is there any wonder then that science properly interpreted supports the Biblical revelations given us by Moses and the prophets? Can proper science be taught in the schools that doesn't support Scripture? The answer in most cases will be no, because you cannot separate "fact" from "truth."

If we thus arm ourselves with the knowledge that prior to the beginning, God was pre-existent, and that it was He who commanded nothing to become something (ex nihilo), and from this command, order was born: *"**For He spoke, and it was done; He commanded, and it stood fast.**"* (Psalms 33:9) Let us further examine His word for more evidence of scientific truth and you will see that the only science that makes real sense as to how the world was and is, is the science called "Creation Science."

The first description the Genesis account gives us of the primeval earth is found in verse two. God was calling into molecular substance those atoms He first created. Scripture then tells us that *"**The earth was without form and void....**"* (v.2), the earth as it was at this point, didn't have any definite shape or form, but simply free atoms which were beginning to collate into molecules. Eventually these molecules formed into a huge sphere of water which God would ultimately use to make up the entire universe. The second half of Genesis 1:2 says: *"**And the Spirit of God was hovering over the face of the waters.**"* It is quite apparent (according to Scripture), that the earth, and as I just mentioned, the universe in its entirety would be formed from this huge sphere of water which God called into existence. Dr. Humpherys estimates that this sphere of water had a radius of about one light year.[1] (6 trillion miles)

In its initial stages, prior to its becoming water, it was a huge cloud of gas. Countless trillions of atoms were flowing faster and faster into this huge cloud until it grew to the exact size God wanted it, then He gave the command: *"**Let there be light.**"* At this instant, the cloud began to collate or condense into the huge sphere of water just mentioned. Atoms began to combine together according to their valence. Inert gases formed in their place, and other compositions were coming into existence. Crushed under the weight of forming molecular substance, the atoms exploded into brilliant light...The beginning of God's primeval universe was bathed in the light of its own glowing atoms. Then God gave light eternal power over darkness and a division occurred forever!

On day two, God caused an expansion of the rest of the waters into outer space. As mentioned, this water would be the building material for the rest of the universe. We find this in verses 6 and 7, which say: *"**Then God said, 'Let there be a firmament in the midst of the waters, and let it divide the waters from the waters.' Thus God made the firmament, and divided the waters which were under the firmament from the waters which were above the firmament; and it was so.**"* The Hebrew word for *"firmament"* is *"raqia"* which means "expanse," or "space." Thus, a greater portion of the water was expanded outward into the universe.

Keep in mind that the earth, at this point, is still a ball of water. The next creative act God did (apparently) was to draw a portion of this water from off the face of the earth and place it in the outer portion of the atmosphere. Thus, the earth had a water vapor canopy which surrounded it at an altitude of about 140 miles high.

The tremendous weight of the material earth pressing toward its center created intense heat and a molten core was established.

The next creative act (v.9 & 10) was to separate the water from the dry land: *"**And God said, 'Let the waters under the heavens be gathered together into one place, and let the dry land appear'; and it was so. And God called the dry land Earth, and the gathering together of the waters He called Seas. And God saw that it was good.**"*

The earth then, we are told, had a water vapor envelope which surrounded it (see Genesis 7:11 and Proverbs 30:4). It had one ocean and one large land mass. It is my belief that the early earth was comprised of two thirds dry land and one third water (ocean). Today, of course, this situation is reversed.

Then God said: *"Let the earth bring forth grass, the herb that yields seed, and the fruit tree that yields fruit according to its kind, whose seed is in itself, on the earth, and it was so. And the earth brought forth grass, the herb that yields seed according to its kind, and the tree that yields fruit, whose seed is in itself according to its kind. And God saw that it was good."* (Genesis 1:11,12)

Notice that the earth was commanded to *"bring forth"* the world of botany. This is exactly what scientists maintain. All the sciences properly interpreted indicate that life did, in fact, spring forth from the earth. Creationists have no argument with this type of teaching. The argument is, and must be, how was life able to arise from non-life?

The next creative act was to place the luminaries in space (Genesis 1:14-22), the fish in the ocean and the birds in the sky. This was accomplished on the fourth and fifth days of creation.

There are several interesting sentences found in the creation account where God established the laws of genetics in both flora and fauna. In the world of flora (verse 11) and again in the world of fauna (verses 21 and 24), we read these words: *"Each after its own kind."* That is, each species would produce progeny only after its own kind. While volumes have been written in the field of biology and genetics, Dr. Harry Rimmer, in his book titled *"The Theory of Evolution and the Facts of Science"* wrote: *"....the greatest sentence ever written in biology are the words 'each after its own kind."* [2]

As we look further into the laws that God established, we will once again find that the written Word of God, says what it means and means what it says. The science upholds the Scripture, or does the Scripture uphold the science?

Evolutionists themselves readily admit the truth of the Biblical record where the fixity of the species is concerned: *"Kinds of organisms have always been observed to produce like kinds, without exception...."* [3] And, *"....only varieties within a species do interchange genetic materials and no new species have been detected."* [4]

While it can be argued that some animals do interbreed such as the horse and donkey, it should also be pointed out that a horse and a donkey are members of the same family. It can also be pointed out that the offspring of such a union, in this case a mule, cannot reproduce.

Animal species then, according to God, are fixed! Animal species according to scientific observation and experimentation are fixed and there are *"no new species being detected."* If evolution were true, the Bible would be wrong and the scientific evidence would support the evolution model, however, by their own words, they have been unable to detect the occurrence of any new animal kinds.

In the book *"Ape Into Man"*, we find this amazing statement: *"Man and chimpanzees proved to be as close as sheep and goat, two species that had always been regarded as very close. Man and chimpanzee are more closely related than horse and donkey, cape buffalo and water buffalo, cat and lion, or dog and fox. Thus we are much nearer to the chimpanzees than to the monkeys many had believed to be our immediate ancestors."*[5]

On what scientific observation or experiment do they base such a statement? None, therefore it is not science but an assumption or an opinion.

Let's follow this teaching to its logical conclusion. You have already learned that the laws of genetics prohibit interbreeding of different animal classes. The law however is flexible enough to allow some close relatives to interbreed and we know that through observation. The book says that *"man and chimpanzee are more closely related than horse and donkey."* The question is: can a horse and donkey interbreed? The answer of course is yes, because they are closely related. Can a man and a chimp interbreed? The answer to that is no....there is no exception, never has been, never will be. Why? Because of the immutable law that God set forth. Because of the greatest sentence ever written in biology: **"....each after its own kind...."** Man and chimp are not related, the Bible declares it, and science upholds it.

Just because some animal kinds are quite similar in appearance doesn't imply a relationship. Michael Denton, molecular biologist, medical doctor, and author, confirms this statement: *"Then there is the problem of convergence. Nature abounds in examples of convergence....In all the above cases the similarities, although very striking; do not imply any close biological relationship."*[6]

"....it is now clear that the pride with which it was assumed that the inheritance of homologous structures from a common ancestor explained homology was misplaced; for such inheritance cannot be ascribed to the identity of

genes. The attempt to find homologous genes, except in closely related species, has been given up as hopeless."[7] (The word "homologous" means to have a resemblance to another biological life form or, having a similar appendage.)

References

1. D. Russel Humpherys, "Starlight and Time" 1994, p.70
2. H. Rimmer, "The Theory of Evolution and the Facts of Science" p.59
3. George Lindsey, "Evolution - Useful or Useless" (Acts And Facts, Oct.. 1985), #148
4. E. Mayr, "Animal Species and Evolution (University Press), p.448
5. S. Washburn & R. Moore, "Ape Into Man" (Univ. of Calif. Berkley, 1974), p.13
6. M. Denton, "Evolution A Theory In Crises" p.178
7. G. de Beer, "Homology: An Unsolved Problem" p.15

Notes

SHOW ME!

Webster's Dictionary defines science this way: *"Knowledge, especially of facts or principles gained by systematic study; a particular branch of knowledge, especially one dealing with a body of facts or truths systematically arranged and <u>showing the operation of general laws.</u>"* (underline added)

While we are all biased toward different ideas and areas of study, we should at least be objective about those biases; otherwise we truly become "the blind leading the blind." Scientists are human and they are biased toward an idea or hypothesis also, particularly when that idea or hypothesis runs against their own brand of science or thought. An example of such bias is put forth in a letter to Dr. Carl Baugh from the professor of geology at Penn State University. Professor Hinderliter wrote:

"....I would have to say that the belief in evolution is in a state of terminal illness. But its death will only be admitted by a new generation of scientists whose minds have not been prejudiced by the type of education now prevalent in the nation's public schools - an education which starts with the belief that evolution happened, which interprets all evidence according to that faith, and which simply discards any evidence which cannot be fitted into the evolutionary framework."[1]

I used to believe in evolution and therefore was biased toward that belief. When I became a Christian I became a theistic evolutionist, that is, I believed God used evolution over millions of years to bring about His great plan. Today, I am a "six day creationist." What changed my biased opinion of evolution and millions of years of earth history? Why would I whittle that down to just twelve thousand years or less of earth history and accept as fact six literal days for the creation of everything that exists? After all, this is not the accepted teaching of society and for that matter many churches. It is frowned on by many of the so-called educated and ridiculed by many in our society.

I certainly had no reason to accept six literal days of creation because it was the "in thing to do." Faith, you say! I am ashamed to say that I lacked faith in a literal belief in God's word. I was one of those that believed we needed to help the Bible out by reading between the lines or adding a little to it here and there. Then one day I challenged God: "Show me."

Webster was one of those responsible for changing my biased mind. I want to quote again in part what he says science is: *"....a body of facts or truths systematically arranged and showing the operation of general laws."* Now read that again and let it sink into your mind because I want to demonstrate to you, as God did to me the scientific, literal truth of the greatest book ever written.

While I have been very critical of evolution and demonstrated that there are no "facts or truths" which "show the operation of general laws," the same test must be given creationism. If creation science does not meet the criterion for a sound scientific base, then it must be categorized as a religious ideology and not science!

God said to me one day, "If My word is not all truth, then what man among you is smart enough to divide the literal truth from the untruth?" I said, "OK Lord, if your word is all truth, show me!"

I had just finished re-reading Genesis 7:11 for the third time. "How can this be?" I muttered under my breath. "Lord, I have a problem!" I exclaimed out loud, "Your Bible seems to indicate that a great deal of water was stored in the outer portion of the earth's atmosphere!" "Excuse me Lord for being so naive, but what held that water up there? And how did the sun shine through it, much less the stars and the moon which is why You made them; 'to give light' on the earth at night. Lord, I'm sure somebody made a mistake in translating Your Word didn't they?" Maybe in the original Hebrew it says something else, I remember thinking to myself.

Since that conversation with God I have received hundreds of hours of schooling in the field of meteorology and climatology. Weather was an important part of my job as a "Fire Management Specialist" for the South Dakota Division of Forestry. I was involved with all areas of fire management and the fire related eco-systems of the western conifer forests, which are not only fire tolerant, but are in fact, dependent on fire for their very survival and health.

As I was studying meteorological laws one day I read: *"A major problem for the atmospheric scientist concerns the abundance of water vapor in the* upper stratosphere."² and, *"Investigations carried out by rockets fired to great altitudes turned up evidence that water cluster ions are found at 185 kilometers in the "D" layer of the atmosphere."*³ As I finished reading that, the words reverberated through my mind, "Show me Lord."

I was awed; a portion of that water is still up there! At an altitude of between 110 and 140 miles, the temperature ranges between 200° f and 240° f. All the water in that zone would be held in an invisible state of translucency by the high temperature. Water is lighter in its vaporous state than dry air. It rises and is held in place as an invisible gas by virtue of its high temperature.

How much water had been up there in the beginning? I can't answer that question, however, in his book titled *"The Waters Above,"* author, Joseph Dillow produces a mathematical model which indicates there to have been about 40 feet of precipitable water. Whatever was up there, it was dense enough to burn up all meteorites before they reached the earth's surface. Geologists have struggled with this problem for years. There has never been found in preflood strata any type of meteoritic material. There seems to be no plausible reason for this absence, and an explanation thus far escapes them. Could this problem be answered by the knowledge of an enriched atmosphere of water vapor, placed at the right altitude by an all knowing Creator?

*"In older geologic formations, no signs whatsoever of the presence of meteorites have been found."*⁴ Along a similar vein, a geologist by the name of Brian Mason agrees with Heide: *"The possible occurrence of meteorites in older geological formations has been a matter of considerable controversy. There seems to be no valid reason to suspect that meteorites have not fallen through geological time. Nevertheless, it has been remarked that no fossil meteorites have ever been found."*⁵

Another variety of meteorite - tektites, (a glassy type of meteorite) reveals a similar pattern. None are found in pre-flood strata: *"Neither tektites nor other meteorites have been found in any of the ancient geological formations."*⁶

How many times in the recent history of this great nation have we failed to correctly analyze scientific data simply because we are so biased toward a theory? We fail to analyze what the reason is for a lack of meteorites in pre-flood strata because of our bias toward evolution. Likewise recent finds in the Paluxy River bed near Glen Rose, Texas, (and more recently, several areas in Russia) have turned up dinosaur tracks in the same stratum as human foot prints.

Evolutionists tell us this is impossible as dinosaurs and man were never contemporaneous. Yet, in the same rock formations Dr. Carl Baugh's team of geologists continue to unearth not only dinosaur tracks and human tracks, but also human artifacts! These finds should be declared as some of the greatest archeological and anthropological finds in history, yet they are hardly mentioned. Why? Simply because these finds agree with the Biblical concept of creation. *"If these findings were not related to the Bible, no one would think of challenging them. The excavations have been conducted properly, and the conclusions should be taken on their merits."*[7]

One can only speculate as to the total effects a vapor canopy would have on plant and animal life; however there are some known factors which we can be fairly certain of.

As has been mentioned off and on in this writing, any scientific theory should operate within the confines of the physical laws which govern our planet. If such a condition such as a vapor canopy existed as recorded in the Scripture, there should be an abundance of evidence to support the story. I have already discussed with you two fairly good sources of evidence to support the vapor canopy theory, but let's look into this a little further.

Meteorological events, like any other operable system on this planet, are governed by known physical laws. It is important that you understand some very basic concepts to a fairly complex science. This science is called meteorology, and as you probably know, is the study dealing with weather and other atmospheric phenomena. By applying the known laws of meteorology to the vapor canopy we can conclude with reasonable accuracy the effects the canopy had on our planet. Let's first discuss a common event: rain. In order for there to be rain, several things must happen. The moisture in the air must be lifted, or pushed up, so it can form into clouds. There are three ways this lifting occurs.

1. The air mass is lifted oreographically, or,
2. it is lifted by thermal convection (heating), or
3. it is lifted by approaching frontal systems (cold fronts).

In conjunction with one of these three things, there also must be nuclei for condensation (small minute particles made up of either dust, smoke particulates, sea salt crystals, plant pollen and even spores). Water vapor which is carried aloft, can form around the small particles just mentioned.

The third event that needs to occur to our water vapor is that it must be cooled so it can condense into larger droplets. This cooling process is brought about by the adiabatic lapse rate; that is, with an increase in elevation there will be a decrease in the ambient air temperature. This rate of temperature decline with an increase in elevation usually runs between 3° f and 5° f for every 1000 ft. of elevation rise.

Moisture in the air is lifted oreographically when wind blowing vapor laden air, runs into a mountain range or a group of hills (fig.1). The result of wind contacting the earth's topographical features will be a lifting of the air mass and its subsequent water vapor into the sky where it is cooled and formed into clouds. This is the reason you always see clouds hanging over the tops of high mountain peaks.

If there is sufficient amount of nuclei for condensation present in the air, and enough water vapor being lifted, then the air will reach its saturation point and rain will result.

An air mass is lifted thermally as it heats up during the day. If there is sufficient water vapor present in the rising air mass, it will cool at altitude, mix with some form of particulate matter and rain will result. It is important to understand that moist air is lighter than dry air at the same temperature. Thus, moisture laden air will more readily raise then will dry air. Then if we add a certain amount of thermal heat to this moist (humid) air mass, it will be thrust upward with enough force to build a convective thunder cell. This upward movement of the air mass is called "convection," or "instability."

Frontal lifting is accomplished when a cold front slides in under warm moist air.

Show Me! 83

Since the cooler dryer air is heavier, it forces the warm air to rise. (fig. 3). A cold front on the surface of the earth would look like a bubble skimming across the water on a pond. The warm moist air in advance of the approaching cold front is pushed up by the leading edge of the frontal system. This lifting of the warm air mass produces clouds at altitude and thunderstorms usually develop just ahead of the approaching cold front.

References

1. Letter from Professor Hilton Hinderliter to Dr. Carl Baugh, from the book "Dinosaur" by Carl Baugh.
2. W. Webb, "Structure of the Stratosphere and Mesosphere" p.129,130
3. E. Ferguson, "Journal of Geophysical Research" p.74
4. F. Heide, "Meteorites" p.32
5. B. Mason, "Meteorites" p.4
6. R. Stair, "Tektites and the Lost Planet" (Scientific Monthly), p.11
7. Baugh, "Dinosaur" p.47

GOD'S TERRARIUM

The thought of an invisible blanket of water vapor surrounding the earth began to excite me. This blanket of water, being held in suspension by a temperature of 200+ degrees, would create a terrarium effect on the earth. A warm even temperature throughout the world would, in fact, make the earth like a "garden." The implications of this idea were almost endless. A warm even temperature around the globe would mean tropical vegetation at all latitudes with the exception of the extreme north and south poles. It would mean luxurious vegetation growing to gigantic sizes in lavish abundance.

A "garden," I thought, the whole world would be like a "garden." The weather would always be perfect. No rainy, stormy weather to ruin picnics or ball-games! In fact, there would only be a few scattered clouds in the morning and again in the evening. As water vapor evaporated from the earth and rose into the sky, it wouldn't condense because the temperature would remain constant as the altitude increased, i.e., no adiabatic lapse rate!

There would be no wind as there would be no cold fronts pushing into warm air masses, nor would there be any dust or sea salt crystals being blown aloft to provide a nuclei for condensation. No rain, no wind or stormy weather, the earth was the perfect place to live!

You may be wondering how the world of botany was watered without rain. My answer to that is just as plants are watered in your terrarium! You pour water in the terrarium and the roots of the plants absorb the water from the soil and release water back into the air along with oxygen through the leaves. In the evening, when the temperature cools a little, the moisture that had risen from the plants and soil during the day would collect on the leaves as dew. This watering cycle is completed every night and day. The hydrological function of the earth in its original form would have operated in the same manner, an ideal botanical system built by an all-knowing Creator!

I know that the laws of meteorology would prevent rain from falling, as I've just explained to you. But does the Bible say anything about the meteorological conditions during that period of earth's history? Once again I found my ancient text upholding modern science as I read: *"....for the Lord God had not caused it to rain upon the earth...but a mist went up from the earth and watered the whole face of the ground."* (Genesis 2:5,6)

There are other evidences which support the vapor canopy teaching of Genesis, and we will examine some of these.

The fossil record which evolutionists used to boldly claim supported their Darwinian theories, is now the very thing which is being used by creation schooled geologists to dispute the evolution model, and dispute evolution it does. Not only is evolution disputed from a scientific standpoint, the creation model of origins is being placed on firm ground. Eyes are being opened to the truth of God's eternal word, and creationists are, for the first time in 60 years, boldly taking the offense.

I have stated before that if these things have happened, there should be some kind of tangible evidence with which to work. If the earth were like a terrarium or a tropical garden, couldn't that be shown in the fossil record? My answer to that is yes. Let's see what the fossil record tells us of the Genesis account, not from the findings of a creation scientist, but from the findings of the evolutionists themselves.

In his book of geology, Clifford Simak writes of what he finds in the fossil record and its testimony of a world that no longer exists: *"Trees, tree ferns, cordaites and scale trees bore great crops of succulent leaves in luxurious profusion, which today is found only in the tropics. The leaves were large and their texture bears witness to the fact that they enjoyed rapid growth. The absence of growth rings also indicates that there was no interruption of the growing cycle."*[1]

It's interesting that many tree species had no *"growth rings"* and apparently grew continually all year long! This is exactly what we would expect if there were in place around this planet a vapor canopy. There would be no winter or even fall, but twelve months of summer and because of this uninterrupted growing season, all plants would become gigantic in size and many would grow year round.

Let's take a look at fossil insects. Were they also larger and healthier? *"The most remarkable feature about such fossil insects as are known is that they are very similar to those living now. In many cases, however, they are much larger than their modern relatives. There are giant dragonflies, giant cockroaches, giant ants and so on. But their form is no different in essence from that of modern insects."*[2] and, *"Fossil dragonflies are well known, the largest of which have a wing spread of eighteen inches."*[3]

This condition is not only true of plants and insects, but <u>all</u> forms of life that lived on earth in those days. *"The modern elephant found in both Africa and India is dwarfed by the fossil imperial elephant which in turn was smaller than the hairy mammoths or the wooly mastodons. Ostriches have been found in fossil form; their size is comparable to the modern giraffe. Fossil pigs have been found as large as rhinoceroses. Fossil slothes have been found, some 18 feet tall, and bears, some 20 feet tall. Some fossil bears, by comparison, make the Alaskan Kodiak bears appear as dwarfs...the same principle holds true for birds, insects and reptiles. Birds with 20 and 30 foot wingspans are known in the fossil record...Reptiles 75 and 85 feet long, and weighing 40 and 50 tons are also known in the fossil record. The remains of beaver have been found and testify of such animals weighing in excess of 600 lbs."*[4]

It should be obvious to anyone that back in the earth's early years, things were considerably different then, than they are today. Evolutionists are willingly ignorant of the conditions that prevailed on the earth to produce such a fantastic array of fauna and flora. Peter alludes to this intentional ignorance and he succinctly wrote: **"For this they willfully forget: that by the word of God the heavens were of old, and the earth standing out of water and in the water, by which the world that then existed perished, being flooded with water."** (II Peter 3:5,6) Indeed, the very thing that produced the fossils these men are writing about was the flood that only Noah and those aboard the ark survived!

One of the benefits most quickly realized from a water canopy would be an atmosphere richer in oxygen content. And, while the water vapor envelope was in the form of translucent gas, it would still exert tremendous pressure on the atmosphere beneath it. Just as a thunder cloud can hold tons of water vapor, so also would this canopy which completely encircled the earth. It has been estimated that this canopy produced an atmospheric pressure of approximately 30 PSI (J. Dillow, *"The Waters Above"*). Thirty pounds per square inch of pressure is more than twice the pressure of today's atmosphere which has an average of 14.7 PSI at sea level.

The botanical world would also benefit greatly from the vapor canopy because the atmosphere would have undoubtedly been far richer in carbon dioxide content then today's atmosphere. If the CO_2 content of the early atmosphere were triple that of today's, the response of the vegetative community would also be tripled. It can be shown that the efficiency with which plants use water to produce organic matter essentially would triple if the CO_2 content were tripled. Because plants use water more efficiently with higher CO_2 concentrations they would also have been able to grow in areas where today they could never thrive.

An atmosphere rich in CO_2 would increase the plant's rate of metabolism, which equates directly to the production of more oxygen. This is because the leaves, exposed to the sun would have an increased rate of photosynthesis, thus, more growth.

At more than twice the oxygen content, all mammals would live healthier and longer lives. The fossil record seems to bear this out. People and animals would live longer for several reasons: *"Unquestionably, hyperbaric oxygenation can often reverse the side effects of aging."*[5]

Probably the best explanation for the longevity of people and animals would be the absence of ultraviolet radiation from the sun. To help you better understand how ultraviolet light is filtered by our atmosphere, let me explain a little further.

"In the upper atmosphere, the solar wind beats down unrestrained upon the atmosphere until the solar rays, especially of the ultraviolet frequency, strike the oxygen molecules and ricochet. When they strike the nucleus of the oxygen atom in just the right way, they will split the oxygen molecules (0^2).

The free oxygen atoms immediately recombine into another form of oxygen known as ozone (O^3).

"The Earth's shield of ozone is a canopy which is invisible to the human eye. The ozone canopy is a vital factor in the composition of our outer atmosphere. If it did not occur, and the short-wave rays of the sun could beat down directly upon the Earth's crust, there would be a sudden extinction of all life processes on the Earth's surface."[7]

There is little doubt in anyone's mind about the importance of ozone. It is known, that if all the ozone in our atmosphere were condensed today, it would fill a container about the size of a shoe box. This layer of atmospheric ozone, if it were compressed, would only be 1/16th of an inch thick. But, what would happen if there was more ozone in Earth's ancient atmosphere? If this ozone was so thick that it filtered out most of the UV light, wouldn't that also add to the longevity of all the biological life forms on earth's surface? Dr. Austin Brues, Director of the Argonne National Laboratory writes: *".... experiments have shown that a single dose of radiation which does not kill an animal within the period of acute radiation sickness may tend to shorten life. Studies using radiation may lead to an understanding of this most least understood fact about life, the aging process."*[8]

In writing on this same subject, George Beadle, Nobel Prize winner for his work in biochemical genetics and head of the Biology Department at California Institute of Technology wrote: *"In experimental animals, the mouse, for example, sub-lethal doses of radiation appreciably reduce the life span. It is almost certain that this also occurs in man. Most investigators agree that there is no threshold below which ionizing radiation has no effect on living matter."*[9]

There seems to be fairly sound scientific verification that radiation in any amount, can, and does shorten our life-span. I think that it could also be shown that longevity would be greatly enhanced if there were no radiation at all reaching the earth's surface.

It is my conclusion that the level of actinic radiation (a term which includes all forms of short wave radiation) at the earth's surface was zero! The reason for this is due to the vapor canopy. This canopy worked two ways. First, it helped to filter the sun's short-wave radiation. Keep in mind that it is these rays which, upon striking the oxygen molecule, break them down into single oxygen atoms. As I mentioned earlier, these single atoms re-combine into groups of three which makes up the ozone molecule (0^3). Some of the sun's rays, upon reaching the earth's surface, are re-radiated back into outer space; these re-radiated rays are known as "long wave-radiation."

Long-wave radiation changes the ozone back into its original state, that is, oxygen. *"But the water vapor canopy not only shielded the Earth from solar radiation; it similarly shielded the outer ozone layer from the Earth's long-wave radiation. It was a buffer zone in the atmosphere. The Earth's long-wave radiation is what causes the ozone (O^3) to recombine back to its normal diatomic state of oxygen (O^2). A reduction in the Earth's long-wave radiation at upper atmosphere levels, absorbed by the intervening water vapor canopy, possibly allowed for an even thicker ozone canopy than exists in the present age. Thus the ozone canopy more effectively shielded the Earth from the solar wind in the Antediluvian age."*[10] The "antediluvian age," spoken of here was that period of time that existed between the creation of man and the flood of Noah.

If we keep in mind the thought that there was no solar radiation reaching the earth from the sun, and the atmosphere contained twice as much oxygen than it does today; would it not be reasonable to assume that all flora and fauna lived longer and more healthful lives than their modern day counter parts? Let's consider the beaver as a case in point.

You have already read where the fossil remains of beaver weighing in excess of 600 lbs. have been found. It is also interesting to know that a beaver is an animal that continues to grow until the day it dies. I would like to suggest to you that a 600 lb. beaver lived to be quite old! Furthermore, I would also suggest that the dinosaurs and other reptiles grew to their gigantic sizes because they continued to grow until the day they died! They lived very long lives indeed.

How then, can we as seekers of truth, scoff at the long ages the Bible gives to the pre-flood patriarchs? Particularly in light of fossil evidence to support vast ages for animals. If animals lived to be very old, why not people? ***"So all the days of Methuselah were nine hundred and sixty-nine years; and he died."*** (Genesis 5:27)

There is one more animal I would like to discuss before I move on to another subject. That animal is quite popular today but has been extinct now for centuries. This animal is also an important one for helping to determine the truth of our model. That animal, of course is the dinosaur.

Why is the dinosaur so important for either creation or evolution? The answer to that lies in the period of Earth's history when these huge beasts walked the planet in apparent abundance. For if God created all animals in a six day period, six to ten thousand years ago, then man and dinosaur were contemporaneous beings on the planet. On the other hand, if we are products of evolution then the dinosaur had already become extinct millions of years before man came along. Let's check the fossil record again and again you will see evolution is bankrupt of any substantiation.

When anthropologists such as Richard Leaky or Donald Johanson, discover a fossil set of bones or a skull, the scientific world of evolution supported by National Geographic and other humanistic organizations, applaud and blow their brass horns quite loudly. So what if all these "ape-man fossils" turn out to be just....ah, apes! After all, aren't we all entitled to a mistake every now and then?

But! What happens when human footprints are found in solid rock along with dinosaur tracks? "Why, that's impossible, man and dinosaurs were separated by 60 million years, so there is no point in any further investigations!"

In the Paluxy River bed near Glen Rose Texas, a steady stream of dinosaur and human footprints are being uncovered. Even at this writing, new and exciting finds are being unearthed and the theory of evolution is being buried in the rubble of pagan ideologies. How has the established world of evolution responded to these finds? By trying to discredit and belittle the finders: *"On a national television show there was a mock presentation of the team's findings at the Paluxy River. Someone who was supposed to be Carl Baugh staggered on stage with footprints he had supposedly excavated at Glen Rose. When asked what they were he explained that he knew the proper shapes and had chiseled them out of limestone.*

"The audience loved it and a strong prejudice was established against anything under-taken by Carl Baugh."[11] (Dr. Carl Baugh is one of the team leaders doing the excavation on the Paluxy River.)

Time Magazine also ran an article with one intent, to discredit the finds at Paluxy. Under the caption titled: *"Defeat for strict creationists"* (June 30, 1986), Time wrote this story: *"The strictest creationists who take the Biblical story of creation literally, believe that the earth was created only several thousand years ago, complete with all the animal species that have ever lived. In their vision, some species perished in Noah's flood but until then dinosaurs and people walked the earth together.*

"Paleontologists do not argue with those who accept this account on faith. But they take strong issue with practitioners of 'creation science,' who purport to offer scientific evidence that this Fred Flintstone version of prehistory is correct. For decades a strong piece of that 'evidence' has been a cluster of fossilized tracks in the seasonally dry bed of the Paluxy Creek near Glen Rose, Texas. One track of three-toed footprints was obviously made by a dinosaur. The feet that made another track looked human to some. The fact that the two varieties of tracks were made at about the same time, creation scientists have long claimed, shows that humans and dinosaurs coexisted. But thanks to the efforts of investigators like Glen J. Kuban, a computer programmer and amateur track expert - who also happens to believe in the Creator - creation scientists have conceded that the second set of tracks was not human after all."

I would like to suggest that Mr. Kuban stick to programming his computers because since that article appeared in Time, Dr. Baugh's team has found (along with dinosaur tracks) human artifacts buried in stone dated millions of years older than man himself. *"The stone, according to evolutionists, is over 400 million years old, yet the artifact is an iron hammer...How could a man-made object have been made 400 million years ago? And what buried it in sedimentary strata deep in the heart of Texas?"*[12]

Pictured is a hammer that Dr. Carl Baugh found on the Paluxy River while digging for dinosaur tracks. The rock it was embedded in was dated by a laboratory to be more than 300 million years old. Uh, apes didn't even exist 300 million years ago let alone a man capable of making such a tool.

Since that article in Time Magazine appeared, more human footprints have been found and positively identified: *"....at my invitation, members of the forensic staff of a Metroplex crime lab have come out to Glen Rose and taken Dental-Stone casts of those prints.*

"I asked them 'what do you think they are?'

"The answer was immediately given, 'There is no doubt about what they are - they are human footprints."[13]

Scientific experts (not computer programmers), in their own field declared those tracks to be human.

Since the article in Time Magazine, there have been other places in the world where human footprints are showing up along with dinosaur tracks.

In Turkmenistan, for example, dinosaur and human footprints have been known for years. Only recently did a newspaper reporter go to the site and write an article on what he found. Journalist Alexander Bushev reported in the January 31, 1995 edition of the *Komsomolskaya Pravda,* of finding more than 3000 footprints on the Turkmenian plateau.

There is very little doubt that dinosaurs and humans walked the Earth together, this is just as the creation model would predict: *"....there is no doubt about what they are, they are human footprints."*

And how about my Ancient Text, does it even mention the dinosaur? ***"He moves his tail like a cedar, the sinews of his thighs are tightly knit, his bones are like beams of bronze, his ribs like bars of iron."*** (Job 40:17,18)

When God's terrarium disappeared, the dinosaur disappeared shortly after. It has been speculated by creationists that after the vapor canopy was condensed the atmospheric pressure dropped by 50% and so did the level of available oxygen in the air. The dinosaur's lung capacity wasn't large enough to supply enough oxygen for his huge mass and he slowly died out.

There is yet another scenario to the demise of these huge beasts. It is possible that when the vapor canopy collapsed, reducing the atmospheric pressure by half, that an important plant community couldn't handle the 50% loss of pressure and it died out. This in turn might have started a chain reaction in the lower portion of the food chain, which ended with the dinosaur.

One of the more ridiculous theories I've ever read appeared in the Rapid City Journal on October 24, 1991. A group of scientists speculated that:

"Dinosaur flatulence may have helped warm Earth's prehistoric climate..." This article also attributed: *"...extensive volcanic eruptions, rising sea levels and other factors that increased atmospheric carbon dioxide levels. This study also suggests that gas from dinosaurs helped maintain or warm the existing tropical climate during the late Cretaceous when flowering plants and plant eating dinosaurs proliferated the Earth."*

Other theories which have been put forth, are: "Death-rays, giant meteorites impacting the Earth, and comets," however these theories only beg the question. Why didn't it annihilate the rest of the animal kingdom?

References

1. Clifford Simak, "Trilobite, Dinosaur and Man - The Earth Story" p.86
2. Henry Morris, "Scientific Creationism" p.86
3. Harry Rimmer, "The Theory of Evolution and the Facts of Science" p.91
4. Donald Patten, "The Biblical Flood and the Ice Epoch" p.231,232
5. Paul Martin, "Stay Young with Hyperbaric Oxygen" (April 1977), p.28
6. Patten, op. Cit., p.212 W.M. Smart, "The Origin of the Earth" p.56
7. Austin M. Brues, "Somatic Effects of Radiation, Bulletin of the Atomic Scientists, Vol. 14, p.13,14
8. George Beadle, "Ionizing Radiation and the Citizen, Scientific American, Vol. 201, p.224
9. Patten, op. Cit., p.213
10. Carl Baugh, "Dinosaur" p.29
11. Baugh, op. Cit., p.19
12. Baugh, op. Cit., p.141

Notes

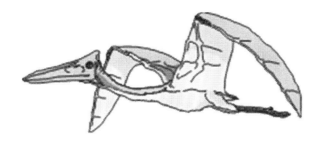

HOW HIGH'S THE WATER MAMA?

"There is no scientific evidence of a global flood. Flood theories have four critical failings: Where did the water come from; where did it go; the ark couldn't possibly hold all living animals, much less the now extinct ones; failure to explain the fossil record. For a global flood to cover Mt. Everest, 5.5 miles of water is required."

This article (in part) appeared in the Rapid City Journal on June 29, 1986. It was put forward as a scientific argument against the Biblical account of a worldwide deluge.

It reflects the ignorance of evolutionists toward creation science. However, the sad part of this article is the number of Christians who also will reject the Biblical account of Noah's flood as factual because of articles written by ignorant men in the name of science.

In these next chapters I will briefly discuss the Biblical flood, where the water came from, and just as importantly, where it went. For as we hold the theory of evolution to the laws governing our planet; it is now our turn to be under the same scientific scrutiny.

Atmospheric scientists tell us that: *"If all the water in our present atmosphere were precipitated, it would only suffice to cover the ground to an average depth of less than two inches."*[1] But the Bible tells us the highest mountains were covered to a depth of **"fifteen cubits."** (Genesis 7:20). Fifteen cubits is approximately 22 feet. As the newspaper article points out, Mt. Everest is 5.5 miles high! Are we to assume that Mt. Everest was covered with water? My answer to that is yes! If Mt. Everest existed during the flood of Noah it would have had 22 feet of water standing above it. But what if Mt. Everest and every high mountain range that exists in the world today didn't exist until shortly after the flood? Now wouldn't that make a big difference? Before we go to Mt. Everest though, let's see where the water came from and just as important, where it all went.

To be sure, these are no small problems and should one be in possession of an ancient text written by man from his own knowledge, then he and his text would be hopelessly lost in the entanglement of modern science and its dictates. For we understand that the atmosphere could not possibly hold enough water to rain upon the earth for forty days and forty nights until the highest mountains were covered by 22 feet of water. But! I am not in possession of such a book. I am in possession of an ancient book written by the knowledge of the Living God. **"The Lord of Hosts"** is His name, and He is perfect in knowledge and wisdom...and so is His Text!

As we turn to this text we discover that it doesn't simply say that it "rained for forty days and forty nights." In fact, if we take the time to study the text we will discover that it mentions three separate sources for the water which ultimately covered the world. These separate water sources are: **"The fountains of the deep, the windows of heaven, and the rain from heaven."** (Genesis 6:2)

"Fountains of the deep." The Hebrew word for **"deep"** is "thom", which means "subterranean." The crust of the earth was fractured and the waters contained beneath its surface spewed out onto the face of the earth. How much water was stored beneath the earth's crust? To be real specific, an ocean of water was stored beneath the earth's twelve mile thick crust.

If we turn to Job 38:8, we read where God asks Job if he knows *"who shut in the sea with doors?"* Then going to Proverbs we find this passage: *"By His knowledge the depths were broken up...."* (Proverbs 3:20) And then, turning to the book of Psalms we read this amazing verse: *"To Him who stretched out the earth above the waters, for His mercy endures forever."* (Psalms 136:6)

The original crust was *"stretched out"* over the majority of water that now makes up our oceans. In other words, all this water was inside of the earth, beneath the crust. But the original ocean was *"gathered together into one place"* (Genesis 1:9). This ocean was probably a fairly shallow ocean which covered about one third of the earth's surface. It is my contention that the vast majority of water which covered the world during the flood of Noah, came from inside the earth itself when *"the fountains of the deep"* were broken up.

Still pursuing this idea, I find in Psalms 33:7 this verse where God kept an abundance of water, a sea of water in fact: *"He lays up the deep in storehouses."* A storehouse is obviously a place where large amounts of water are being stored. Approximately half of all the water in the world was found here.

What I am presenting to you is not from blind faith, nor as an evolutionist who would ask, even demand, that you accept his ideas solely on faith. But I present to you good physical evidence to support the concept of where most of the water came from.

On the next page is a picture of the Mid-Atlantic Ridge, which is a range of mountains that runs almost from the North Pole to the South Pole. These mountains are called the Mid-Atlantic Ridge because they run beneath what is now the Atlantic Ocean. They are metamorphic in origin and show signs of being altered (hydrothermally) by hot magmatic water. These mountains are thought to be the greatest single mountain system on earth.

As you view the picture, you can clearly see what appears to be a crack at the top of the ridge system. This crack runs the length of the mountain chain which, as I said before, runs from pole to pole. How this crack came to be is unknown...or is it? Let's say that the formation of this crack and the entire mountain range for that matter is unknown to modern science. But as I stated before, I am in possession of the Book of Knowledge and beside it there is no equal.

My ancient text tells us there was a time in the earth's distant past when the crust was *"stretched out"* above a vast reservoir of water (Psalms 136:6). This water was held in an extremely high state of temperature, well into and probably above 1000° f. When the crust of the earth fractured or cracked open, this reservoir, being under tremendous pressure virtually exploded and shot high into the air.

It literally became as a standing fountain that, because of the great pressure, shot up for possibly thousands of feet into the air. This in turn would fall as rain/snow back to earth.

This idea of the earth being cracked open and allowing water to come forth is also upheld scripturally. God's prophet Isaiah wrote these amazing words: *"The earth is violently broken, the earth is split open, the earth is shaken exceedingly."* (Isaiah 24:19)

The Mid-Atlantic Ridge

During the flood epoch, there were tremendous upheavals caused by volcanism, the earth was blanketed with smoke and ash so dense that it brought 24 hour darkness. Earthquakes of a magnitude far beyond our imagination would rend the earth, shaking it from pole to pole. New continents rose from the ocean floor, while others sank to a watery grave.

If the crust were *"stretched out"* over this reservoir of water and suddenly split apart, it would also begin to contract, or pull away from itself on either side of the crack, similar to a rubber band that has been stretched tight, and then cut in the center. This movement of the earth's crust would cause mountain ranges to be pushed up in the coastal regions of the world: *"Mountain masses are raised by orogenic (mountain making) movements. Nearly all great mountain ranges are anticlinoria, that is, the dominant folding is upward at the central parts of the folds. Mountains are generally folded because of strong horizontal movements along the circumference of the earth."*[3]

Over the last four thousand years since the flood, the earth's crust has slowed down to just inches of travel per year. Geologists measuring this slow rate of "continental drift," suppose that at this even uniform rate of movement, it took millions of years for the North American and European continents to drift as far apart as they are today thus making the uniformitarian statement: *"The present is a key to the past,"* irrelevant.

Let's assume for a moment, that you had never seen snow, and you were driven to the bottom of a high mountain peak. You stood there gaping at the beauty of this high mountain which was laden with snow, noticing that the summit was about five miles distant. As you turned your back, looking down slope, a skier whom you hadn't noticed came gliding past you at a very slow rate, finally stopping several feet from where you stood. You walk over to the skier, asking him where he came from. He replies that he just came down from the top of the mountain. You, of course, only saw his run-out speed which was quite slow. You would have no other way of calibrating his speed except by what you saw, therefore you assume incorrectly, that it had to have taken the skier hours to get down from the top.

Gauging history by observing present day events, in many instances, is a very poor way of trying to explain what happened in days gone by. We will see further evidences of this as we get into the next chapter.

As I mentioned earlier, just as a rubber band snaps back when it's cut in the middle, so also, I believe, did the earth's crust move quickly and with great force toward the east (Europe) and the west (North and South America). I think the last part of the verse in Isaiah 24:19 describes just this very thing: *"…. **the earth is shaken exceedingly"**.*

Let's turn now to the source of heavenly water, which I believe was produced by a total collapse of the vapor canopy. The first question one should ask is: "What caused the precipitation of the vapor canopy?" At this point, it would be easy to say: "God did," and leave it at that. However, that wouldn't be science but a proclamation of one's faith. I do know, and believe that God did in fact orchestrate and direct the entire event otherwise the world could have never survived such a cataclysm.

The evidences are quite plain that what the Bible says is true. Example: there was a flood which covered the entire earth. Evidence: the earth is covered with sedimentary rock, laid down according to density in layers. The fossil remains of the abundant life forms on the earth perished in a sudden global catastrophe! *"The general picture of animal history is thus a burst of general body plans followed by extinction."*[4]

A little earlier in this writing, I discussed some of the laws of meteorology. One of the things I discussed was the need for some type of nuclei for water vapor to collect around in order for there to be rain. This nucleus for condensation is provided by (1) sea salt crystals (2) dust (3) smoke particulates. As I mentioned, without cold frontal systems mixing with warm air masses at the earth's surface, there would be very little wind. Sea salt crystals and dust couldn't possibly be lifted high enough to interact with the vapor canopy, at 140 miles above the earth, it was just too high. This leaves us with only one other source for a nuclei: smoke particulates.

Turning in our Bible to Psalms 46:3, we read: *"Though the waters roar and be troubled, though the mountains <u>shake with its swelling</u>."* Then turning to Nahum 1:5, we find this verse: *"The mountains quake before Him, <u>the hills melt, and the earth, heaves</u> at His presence...."* Psalms 144:5: *"Touch the mountains, and they shall smoke."* Psalms 104:32: *"He looks on the earth and it trembles; He touches the hills and they smoke."* Dust? Smoke? Volcanism? Bingo! Volcanism would sure do the trick.

It is my contention that the vapor canopy (***windows of heaven***) were precipitated through volcanic dust and smoke. Furthermore, it is my contention that the dust and smoke blackened sky initiated the so-called ice age. It is also my intent to demonstrate to you the evidence which will support the text Living God. If such an event did actually happen, it should be quite easily checked out. There should be found in the ice at both poles, an abundance of volcanic ash and dust interlaced in the ice sheets which covers both poles. Further, if there were in place around our earth a thermal blanket of water vapor which produced a pressurized atmosphere of 30 PSI, and this thermal blanket suddenly collapsed from the resultant smoke and dust being thrust up into it, it would have produced a sudden and drastic drop in temperatures around the world, but in particular, these temperature drops would be very extreme in the polar regions.

Just as the flood left the world covered with sedimentary rock and fossils, so also should this event have left traces of its existence and subsequent collapse, which would have plunged a warm, tropical to semi-tropical, garden like world, into a frigid nightmare of sudden death and extinction.

The Mid-Atlantic Ridge dividing Iceland

References

1. C.S. Cox, "Water," P.20
2. Emmons, Allison, Stauffer, Thiel, "Geology, P.375
3. Op. Cit., P.375
4. P. Johnson, "Darwin on Trial," P.55

GLACIEOGENESIS

The time of the year was August. The place, what is today known as northern Siberia. It was 1 o'clock in the afternoon and the temperature was a normal 83^0f. The lumbering mammoth cast a wary eye at the large saber-toothed tiger that had been following her and her small calf. The mammoth knew if she allowed her baby to stray very far from her protective side, it would become an easy meal for the hungry predator.

The ground trembled only slightly beneath her feet, and in the distance came a soft rumble, not unlike the sound of distant thunder we hear at the approach of a thunderstorm.

The mammoth was relieved to see the large tiger, obviously startled by something run off. She bent down and gulped up a mouth full of tender sedges and butter cups. As she raised her head, she felt for the first time in her 800 year life a wind beginning to blow. The leaves on the trees began to shake a bit and the grass on which she was standing begins to bend slightly. Even though it wasn't a strong wind she became nervous, something wasn't right. She again looked down at her calf; it was trying to get under her where it could begin sucking.

Suddenly, the wind speed began to rapidly increase. The startled mammoth began to pull away from her sucking calf. The wind was already approaching gale force, trees were breaking off, some were being uprooted, and one smashed down near the terrified mammoth. As the mother turned to her calf, the winds, increasing every second now, swept her 1400 lb. baby away. She tried to follow but was blown over by the power of the wind, now approaching 200 mph. Rain and freezing sleet began to pelt the huge animal. She was numbed by the rapidly dropping temperature and fought to get to her feet. She could not stand in the force of the numbing winds now approaching 200 mph.

Huge chunks of ice driven by 200 mph winds and a rapidly dropping temperature made a quick end of the mammoth and her calf. As the temperature continued to drop ultimately to more than 200⁰f. below zero, the huge beast would be frozen solid in the space of an hour, her destiny to become part of the Siberian permafrost, a well preserved specimen which would be discovered centuries later.

That is my scenario of the beginning of the ice age. Mankind is in possession of only one book that indicates the suddenness of that period of earth's history. There is no other book in the world, modern or ancient, that gives positive indication that the ice age was an abrupt, disastrous event which annihilated all life forms at both poles. Was the coming of the ice age in fact a very sudden catastrophic event as the Bible indicates? Or was it as the evolutionists believe, a slow gradual process? The scientific facts will be laid out before you. You will be the judge. Modern evolution theory or my Ancient Text? Only one of them can be right.

Evolutionist Archibald Geike, writes of this same period: *"This arctic transformation was not an episode that lasted merely a few seasons and left the land to resume thereafter its ancient aspect.*

"With various successive fluctuations it must have endured for many thousands of years. When it began to disappear it probably faded away as **slowly and imperceptibly as it had advanced.**"[1] (bold print is mine)

Charlies Lyell, the father of uniformitarian thought, also wrote of the ice age: *"Huge amounts of snow originated from water vapor which had risen by the process of evaporation from the earth's oceans and as the vapor moved to the cooler latitudes, it condensed and fell in the form of snow,* **where over a vast span of time,** *it built up enough to cause glaciation. This glaciation then slid south from the northern regions and north from the southern regions, eventually covering large portions of the earth's surface."*[2] (bold print is mine)

Evolution, always of necessity sticking to the slow gradual, uniform processes. Is what we observe today a key to what happened millions of years ago? I think not.

Of the period we are now discussing the Bible says: ***"Out of whose womb came the ice? And the hoary frost of heaven, Who hath gendered it? The waters harden like a stone, and the surface of the deep is frozen?"*** (Job 38:29,30)

In the book of Psalms we read this verse: ***"Praise the Lord from the earth, you great sea creatures and all the depths: fire and hail, snow and clouds: stormy wind, fulfilling His word."*** (Psalm 148:7,8)

In the verse just quoted, the word *"fire"* is followed by the words, *"....hail, snow and clouds: stormy wind...."* are indicative of the collapse of the vapor canopy and the ensuing ice age. As I pointed out earlier, I believe it was volcanic activity that precipitated the canopy. If this is true, then the evidence to support this concept should be found sandwiched in between layers of ice in the Polar Regions.

This is exactly what is found under the polar ice, as we read: *"The frozen muck of the arctic tundra is literally strewn with volcanic material hundreds of feet deep all evidencing a violent volcanic catastrophe."*[3] What about the Antarctic?

"Army cold regions geologist Anthony Gow took over 7,100 ft. of core samples from nine Antarctic glaciers. He found over 2,000 individual volcanic ash falls interbedded with the ice."[4]

Other evidences we find, all indicating the probability of extensive volcanism occurring simultaneously with the ice age are the ice caves of the Pacific Northwest: *"Consider now yet another phenomena, namely ice sandwiched between layers of lava rock. In the intermountain plateau west of the Rockies and east of the Cascades, there was a great outpouring of lava during the flood catastrophe. In some places the lava deposits exceed 8,000 ft. in depth upon original bedrock. At the northern edge of this lava plateau, inflowing ice complicated geophysical features.*

"Great coulees were formed by the rivers of flowing ice. The Columbia Valley, Moses Coulee, and Grand Coulee are several examples. Throughout this area, particularly in northern Washington, but also occasionally in Idaho and Oregon, we find the phenomena of ice caves. Much ice remains

sandwiched in between layers of lava (of igneous origin), and that which has melted has left empty areas, the caves themselves."[5]

As you have just read, the physical evidence indicates the ice and volcanism occurred together. This is evidenced by the ash layered between ice in the Polar Regions and the formation of ice caves in the Pacific Northwest. Great Coulees were formed in a relatively short period of time as the giant sheets of glacial ice slid southward, plowing through water softened sediments.

If there were great winds produced by a carrillous effect and also indicated by the Biblical account, **"stormy wind,"** it should also have left its mark on the pages of geological history. A wind of 200 + mph, with ice and frozen chunks of ash intermingled with it, would literally tear all plant and animal life to shreds! Let's journey now to the northern most reaches of our planet and investigate the permafrost for more evidence to support our ice-age model.

In his book *"Riddle of the Mammoth,"* Ivan Sanderson writes of his bewilderment in what he found in the frozen northland. *"The greatest riddle, however, is when, why and how did all these assorted creatures, and in such absolutely countless numbers, get killed, mashed up and frozen into this horrific indecency."*[6]

Continuing on with this investigation we find: *"In many places the Alaskan muck is packed with animal bones and debris in train-load lots. Bones of mammoth, mastodon, several kinds of bison, horses, wolves, bears, and lions tell a story of a faunal population. The Alaskan muck is like a fine, dark gray sand...within this mass frozen solid, lie the twisted parts of animals and trees intermingled with lenses of ice and layers of peat and mosses. It looks as though in the midst of some cataclysmic catastrophe of ten thousand years ago, the whole Alaskan world of living animals and plants was suddenly frozen in midmotion in a grim charade."*[7]

As for the evolutionist teaching that the ice age *"probably faded away as slowly and imperceptibly as it had advanced...."* let's zero in on the physical scientific facts. Facts which, as you will see show evolution for what it really is. A pagan assumption, bankrupt and void of any scientific evidence. *"The present"* is not a *"key to the past."*

We will now shift our attention to the waste-lands of Siberia where our frozen mammoth has just been discovered.

"The lessons of the animal are striking. At first glance, it appears that a large animal was peacefully grazing on buttercup flowers and was suddenly overtaken by a deep freeze in the middle of summer. The plant remains in the stomach of the Beresovka mammoth indicate that the animal died in late July or early August. Furthermore, the animal froze quickly enough to leave these stomach contents in a well-preserved state and for at least some of the meat on the carcass to be edible."[8]

Continuing on with this discussion of the Beresovka mammoth, Dillow writes: *"The mammoth must have been over-whelmed suddenly with a rapid deep freeze and instant death. The sudden death is proved by the unchewed bean pods still containing the beans that were found between its teeth, and the deep freeze is suggested by the well-preserved state of the stomach contents and the presence of edible meat."*[9]

While examining the vegetation in the stomach of the Beresovka mammoth, Russian physiologist Victor Sukachev found the plants extremely well preserved. Again, Dillow writes: *"What precisely did Sukachev find? He discovered that the blossoms of the Alopecurus alpinus in the stomach were so well preserved that he could establish the species with exactitude. Apparently, in the case of the Alopecurus alpinus, the delicate hair-like follicles on the leaves were so well preserved that Sukachev could relate them to a particular species. Even the color of the leaves - brown, was still intact, indicating that no leaching of the pigment occurred prior to freezing."*[10]

In laboratory experiments, butter-cup flowers were immersed in the stomach fluids of an elephant. Within 45 minutes the flowers were disintegrated so badly as to be unrecognizable! The evidence therefore is unmistakable; the Beresovka mammoth was frozen solid within a period of 45 minutes! Keep in mind, that a mammoth was twice as large as a modern day elephant. The best estimate of the temperatures required to freeze an animal this large, this quickly, is a minimum of $230°$ f below zero, with a strong wind.

Hopping back across the ocean from Siberia to Alaska, another geologist records his puzzling find: *"We have already seen that the muck pits of Alaska are filled with the evidence of universal death. Mingled in these frozen masses are the remains of many thousands of animals killed in their prime. The best evidence that we could have that this Pleistocene death was not simply a case of the bison and mammoth dying after their normal span of years is found in the Alaskan muck. In this dark gray frozen stuff is preserved, quite commonly, fragments of ligaments, skin, hair, and even flesh. We have gained from the muck pits of the Yukon Valley a picture of quick extinction. The evidences of violence there are as obvious as the horror camps of Germany. Such piles of bodies of animals or men simply do not occur by any ordinary natural means."*[11]

What about the vegetation of the pre-flood world? Is there any evidence other than the fossil record which bears ample proof, that the entire world was tropical to sub-tropical and like a *"garden?"* In his book titled *"The Energy Non-Crises"* author, Lindsey Williams writes: *"There is an interesting point to mention in passing. Though the ground is frozen for 1,900 ft. down from the surface at Prudhoe Bay, everywhere the oil companies drilled around this area they discovered an ancient tropical forest. It was in frozen state, not in petrified state. It is between 1,100 and 1,700 feet down. There are palm trees, pine trees, and tropical foliage in great profusion. In fact, they found them lapped all over each other; just as though they had fallen in that position."*[12]

Tropical vegetation in northern Alaska? Huge animals frozen solid within minutes? And how does one explain tropical vegetation being buried under 1,700 feet of sediments? Ask yourself this question; is the present really a key to the past? Does the evidence support a slow gradual encroachment of cold weather as the evolutionists say? This is the wisdom of the world, and as you can see, it doesn't fit the scientific evidence!

But my Ancient Text, now here is a book. A book where knowledge can be found if a man will set his heart to know and understand knowledge. Job 28:27 says: **"Then He saw wisdom and declared it; He prepared it, indeed, he searched it out."** This *"wisdom"* which God has *"declared,"* you have just studied in part. It's the infallible Word of God.

The Bible says the world was as a *"garden,"* and you have seen the evidence.

The Bible says there was a worldwide flood and almost all life perished. You have seen the evidence.

Finally, concerning the coming of the ice age, the Bible declares: **"From the breath of God ice is made, and the expanse of the waters is frozen. He casts forth His ice as fragments, who can stand before His cold? Praise the Lord from the earth you great sea creatures and all the depths; fire and hail, snow and clouds, stormy wind fulfilling His word."** (Job 37:10, Psalms 147:17 & 148:7,8)

I have put together a sound expose` for the problem of where the water went, I want to give you a good concept to answer the question that some ask, "how could the earth have been dried off in the space of about a year? After all, if the sun came out and evaporated the water, wouldn't it just make clouds and rain again?

The answer to the question, of course is yes, it would have made clouds and more rain. But remember, it was cold and most of the world's precipitation would fall in the form of snow. This snow would be transported north of the equator to the northern regions of the globe. Hence, the build-up of glacial ice sheets on the northern continents. The same process would be happening in the southern hemisphere. In this way, most of the water was stored in the form of snow and ice.

How did Noah and his family take care of all the animals on the ark? What do many animals usually do when it turns cold? They hibernate! I believe that most of the animals hibernated and Noah and his family only had to take care of the few animals that didn't hibernate. And since the ark was floating more in the equatorial area of the planet, the water didn't freeze because the average temperatures in that region probably remained around the fifty degree range. But still cool enough for animals to hibernate.

References

1. Archibald Geike, "The New Treasury of Science" p168
2. Charles Lyell, "Principles of Geology"
3. Frank Hibben, "The Lost Americans"
4. Anthony Gow, "Geological Investigations in Antarctica"
5. Donald Patten, "The Biblical Flood and the Ice Epoch" p.120
6. Ivan T. Sanderson, "The Riddle of the Mammoth"
7. Frank Hibben, "The Lost Americans"
8. Joseph Dillow, "The Waters Above" p.321
9. Op. Cit., p.377
10. Op. Cit., p.379
11. Frank Hibben, "The Lost Americans" p.168
12. Lindsey Williams, "The Energy Non-Crises"

Remains of a mammoth

MOUNT EVEREST AND THE FLOOD OF NOAH

The entire world is covered with sedimentary rock. Sedimentary rock is formed when a portion of land mass is covered with water. The faster the water flowed in differing areas generally would necessitate more flora and fauna deposits, and sediment deposits generally would be deeper.

Entire forests could be transported by the powerful hydraulic force of the flowing water. They would be caught and held in place by large tidal pools and buried under thousands of feet of sediments. This process could also transport huge numbers of animals and bury them in mass, in particular places or areas of the globe.

In the case where this flora and fauna happened to be in Polar Regions, they would be preserved by freezing. In the same situation, only moving closer to the equator, the flora and fauna would be preserved by petrifaction or fossilization. The primary difference is the fossilized specimens appear to be older than do the frozen ones so older dates are assigned to them. This also is due in part to the layer of sediments or rock the specimens are found in.

Most fossils have one thing in common, volcanic ash sandwiched in between layers of rock is usually present in and around major fossil beds.

In the science section of Time Magazine, while talking about Australopithecus and baboon fossils I recently read this: *"Last summer Walker, a professor at the John Hopkins University Medical School, was looking for baboon fossils, when he spotted the skull fragments. By studying the volcanic ash and other bones nearby, his colleagues determined the skull's age."* (August 18, 1986)

I believe the volcanic ash mentioned in this article was laid down at the same time as the ash found in the Polar Regions; it's also interesting to note the exact method used to date the skull!

In another article, this one appearing in the Rapid City Journal, titled, *"Ancient Soot Supports idea that a Global Pall Killed Dinosaurs."*

"In a report in the Journal of American Science, published Friday, chemists of the University of Chicago said that 'surprisingly, large amounts of soot appeared to be worldwide, could only have been produced in flames or hot gases and represented fallout from a dense smoke cloud that must have brought a killing darkness and chill to the world....smoke, even more than the dust clouds, absorbed sunlight and sent temperatures plunging worldwide." (Friday October 14, 1985)

Still pursuing this idea, the Rapid City Journal again ran another article about this controversy of what killed most of the animals on the earth. This article ran under the caption: *"Scientists Split over what killed Dinosaurs."* The article states: *"Researchers working on a 65 million-year old mystery are piling up new evidence that comets or volcanoes killed the dinosaurs, with possible help from a death star, acid rain and extended cold and darkness.... Charles Officer, a Dartmouth College geologist, who believes sustained, catastrophic volcanic eruptions, not comets smashing into Earth, spelled mass extinction for up to half of all life on earth.*

"The effects of this intense volcanism would be global cooling, intense acid rain and increased ultraviolet radiation." (December 15, 1985)

I touched earlier on the idea of volcanic action being the primary reason the water envelope collapsed. Volcanic action on a worldwide scale would, as the article indicates, contribute to a drastic cooling of the planet because of dense smoke particulates and dust. It is also a fact that volcanic ash in the upper atmosphere would reflect a lot of sunshine.

The darkness brought about by such extensive volcanic activity is referred to in the Bible as *"thick darkness."* An example of this expression can be found in Deuteronomy 4:11 where it says: ***"Then you came near and stood at the foot of the mountain, and the mountain burned with fire to the midst of heaven, with darkness, cloud, and thick darkness."***

Other examples of volcanic activity are found throughout the Scriptures. Oceanic volcanism: ***"Look, He lifts up the isles as a very little thing."*** (Isaiah 40:15)

Psalms 46:2, 3, says: ***"Though the mountains be carried unto the midst of the sea; though its waters roar and are troubled, though the mountains shake with its swelling."***

These verses are explicitly discussing the pushing up of mountains and islands by volcanic activity.

Not only are mountains raised through the process of volcanic activity, but as you read earlier, mountains are built by: *"strong horizontal movements along the circumference of the earth."*[1] That is, as the crust of the earth began to move, mountains were shoved up, some to great elevations, some considerably less. But before the earth's crust cracked open, and began to move, let's go back to a period of time prior to the flood and examine the earth's surface.

We are told in Genesis 7:20 that the highest mountains were covered by flood waters to a depth of about 20 ft. It is my contention that the highest mountains were not very high at all compared to today's mountains. Geologic research indicates this to be true. *"Many lines of dinosaurs evolved during the 100 million years or more of the Mesozoic history in which they live. In those days the earth had a tropical or sub-tropical climate over much of its land surface and in the widespread tropical lands there was an abundance of lush vegetation. The land was low and there were no high mountains forming physical or climatic barriers."*[2]

"The geologic evidence shows unquestionably that at sometime in our earth's history, the mountain ranges were much smaller and lower than they are today. And these great ranges were pushed up only in fairly recent times."[3]

As the mountains began to rise, the sea floor, by the same token began to sink. Evidence to support this idea can be borne out by the discovery of drowned islands or seamounts: *"In the past decade have been discovered great numbers of seamounts which are nothing but drowned islands out in the middle of the ocean. These are flat-topped, and therefore non-volcanic in formation and are now in many cases more than 1,000 fathoms (6000 ft) below the surface. Yet they give abundant evidence of having once been above the surface."*[4]

It can also be shown that the sea floor has indeed sunk down, in many places thousands of feet: *"Can we, as seekers after truth, shut our eyes any longer to the obvious fact that large areas of sea floor have sunk vertical distances measured in miles."*[5]

As the earth's crust fractured and buckled, it not only moved horizontally but also vertically in many places, pushing up large mountain ranges as tectonic plates were thrust over the top of other plates.

Paleontologists in discussing the dinosaur dying out, give these reasons as to his demise: *"....dinosaurs were already on the wane, and may have even died out, because of drastic changes in the climate from falling sea levels, separating continents, volcanic eruptions and emerging mountains."*[6]

You have read some of the scientific evidences as to how the great mountain ranges came to be. You have also read about the sinking of the sea floor, in many instances the sea floor has: *"sunk vertical distances measured in miles."*

Turning now to the greatest book ever written; the Book of books which has already, by the wisdom of God declared these things thousands of years ago:

"You laid the foundations of the earth that it should not be moved forever. You covered it with the deep as with a garment; The waters stood above the mountains. At Your rebuke they fled: At the voice of your thunder they hasted away. The mountains rose, the valleys sank down to the place which You appointed for them." (Psalms 104:5 – 8 Amplified) **"The mountains rose, the valleys sank down...."** It is my contention that Mt. Everest or any of the high mountain ranges of the world didn't exist until shortly after the flood, as these mountains show signs of being" *"pushed up only in fairly recent times."*

Now I ask you, does modern science when properly interpreted support the Scriptural Text? Clearly it does. But that's because the One who made it all happen was also faithful to record it for us; that we can see Him through His word.

When the ark finally came to rest on the mountains of Ararat, (Genesis 8:4), it was still quite cold, particularly in the northern and southern hemispheres. However, in the portion of the world closest to the equator, the temperature probably averaged $55 - 65^0f$ or possibly higher. But the seas in many places were still frozen (Job 37:10), as the earth was literally held in the grip of an ice age.

Nonetheless, evaporation took place around the world in the equatorial regions which were considerably warmer, but also in the colder climes, the evaporative process would continue. Large amounts of water were transported through the evaporative procedure. In the northern and southern latitudes where it was cold, precipitation fell as snow where it began to build up into huge sheets of glacial ice.

This accumulation of precipitation in the form of snow and ice did two things. One, the weight of the ice caused the land masses to sink a little, and two, huge amounts of water were stored in the form of ice, eventually to draw down the oceans far enough to begin exposing natural land bridges for people and animals to cross on. Keep in mind that where there were no exposed land bridges to cross on, people and animals may have traversed short sections on ice in those areas where the seas were still frozen.

As people and animals began to spread out, possibly over several centuries, the earth gradually began to warm up, melting the continental ice sheets which on the North American continent were more than a mile thick. And, as these huge reservoirs of water ice began to melt, the continents began to rise as the tremendous load of ice and water were removed.

The melting glacial ice produced large rivers that easily cut canyons through the still soft sediments in their journey to the sea. And, of course, the melting glaciers caused the oceans of the world to begin to raise. (Al Gore's prophecy huh?) The rising of the seas eventually cut off any retreat for people and animals that had migrated to differing regions of the planet.

The world was divided in this manner and new nationalities came into being. And once again, God's faithful word recorded the event: *"To Eber were born two sons: the name of one was Peleg, for in his days the earth was divided...."* (Genesis 10:25). Peleg was born (according to the table of the nations) approximately 400 years after the flood epoch.

In fact, a recent article which appeared in the Rapid City Journal(gotta love that paper huh?): *"Scientist Support Claim of Early Human Arrival in Australia: 'Many scientists theorize that Australia and Indonesia were separated only by a narrow channel of water when sea levels dropped during the ancient ice ages. One of these periods was the most likely time that the ancestors of present day Aborigines came to Australia."* Sept. 25, 1996

For hundreds of years after the flood, the world's weather would fluctuate as it slowly began to warm up. All the water that was being evaporated from off its surface would create a very unstable atmosphere which in turn, would produce numerous mini ice ages and floods.

All the water that covered the earth during the flood epoch of Noah is still here on the earth. This is consistent with the first law of thermodynamics, which if you remember, states that matter can neither be created from nothing, nor can it be destroyed. The water of the flood epoch is accommodated by a deepening of the ocean floors, a pushing up of the land masses, and water storage in the form of ice at the poles where in the Antarctic the ice sheets are still over 10,000 feet in depth.

If the earth were made as smooth as a billiard ball, and the polar ice caps melted, the water would stand more than five miles above its surface!

References

1. Emmons, Allison, Stauffer, Thiel, Geology" p.23
2. E.H. Colbert, "Evolutionary Growth Rates in Dinosaurs" Scientific Monthly Vol. 69, p.71
3. R. E. Flint, "Glacial Geology and the Pleistocene Epoch", p. 514, 515
4. John Whitcomb and Henry Morris, "The Genesis Flood," p. 124,125
5. Author unknown, from the book: "Illogical Geology," Geotimes Vol. III, p.19
6. Rapid City, Journal, "Ancient Soot support Idea that a Global Flood Killed Dinosaurs" October 4, 1995
7. R.E. Flint, Ops. Cit., p.514, 515

Notes

GOD'S WORLD

"For thus says the Lord, Who created the heavens, Who is God, Who formed the earth and made it, Who established it, Who did not create it in vain, Who formed it to be inhabited."(Isaiah 45:18)

Every time I hear the cry of the radical environmentalist, I think of this verse. God did not create the earth *"in vain,"* He *"formed it to be inhabited."* Mankind is so arrogant, so proud and yet so fearful. He actually believes in his heart that he alone holds earth's destiny in the palm of his hand.

If God's Word is true, then we need to understand from the verse just quoted, that He indeed made the earth to be *"inhabited"* and if He made it to be inhabited, then He would also know exactly what we need to survive on it.

In this chapter I want to cover some of the environmental myths which are so prevalent today in our country and even in the public school system.

Environmental myth #1. The ozone myth.

We are destroying Earth's ozone layer by using CFCs! (chlorofluorocarbons) CFCs are a man-made chemical that has been used in aerosols; it is also the major product in the gas Freon that is used in refrigerators and air conditioning systems. An article which just appeared in the Rapid City Journal and titled: *"Ozone holes blamed on man-made chemicals"*—This article states in part: *"New satellite data provide conclusive evidence that annual ozone holes over the South Pole are caused by chlorine from man-made chemicals, not from naturally occurring sources, NASA scientists said Monday."*

There are a lot of "holes" in this idea. First, the gas that is responsible for destroying ozone is heavier than air. How do man-made CFCs get up to where the ozone is? The ozone layer is found between 10 and 20 miles above the earth' surface.

The largest ozone hole occurs annually over the south pole, but the vast majority of the world's industry, or CFC producing nations are located in the northern hemisphere! Also consider for a moment those volcanoes again, they all produce a considerable amount of chloride gasses.

Mount St. Augustine (Alaska) that erupted in 1976 injected 289 billion kilograms of hydrochloric acid directly into the stratosphere. This amount by itself is more than 500 times the total world production of CFCs in a year! It has been estimated that Mt. Pinatubo which erupted in the Philippine Islands in 1991 ejected more chlorine gasses in one year, than mankind can produce in one hundred years. Then we have Mt. Erebus (Antarctica), Mt. Erbus has been erupting constantly for the last 100 years, ejecting more than 1000 tons of chlorine every day. Indeed, the total amount of CFCs produced by man each year is minuscule compared to what nature does on a daily basis.

But let's get back to the real question. Is the ozone in our atmosphere getting thinner? If the ozone is declining on a global scale, then there would be an increase in ultraviolet light reaching the earth's surface. But there is no evidence of such an increase. National Oceanic and Atmospheric Administration (NOAA) scientist John Delouisi said that readings from meters at eight different stations located throughout mainland United States have shown *"an average surface ultraviolet radiation trend of minus 8 percent from 1974 to 1985."*[1] In other words, the amount of ultraviolet radiation reaching the earth has been declining! It's also interesting to note, that two French investigators measured the ozone hole in 1958, when it was thinner than at any time since!

Another NOAA scientist, Melvyn Shapiro, who has spent more than two decades studying how weather affects ozone told Insight Magazine that he sees *"no major alterations in ozone patterns....a storm system moving across the United States on the jet stream, for example, can result in up to a 50 percent reduction in ozone across the weather system."*[2]

Evidently storm systems can and do alter, to some degree the amount of ozone at any given time. These are naturally occurring events and we should not fear them.

There can be no doubt the ozone layer is an important layer of gas in the upper portions of our atmosphere. In fact, all the gasses that make up our atmosphere (depending on the altitude), are found in layers. The Bible has this to say about the layering of the atmosphere: **"He who builds His <u>layers in the sky</u>, and has founded His strata in the earth; Who calls for the waters of the** *sea, and pours them out on the face of the earth - The Lord is His name."* (Amos 9:6)

The production of ozone shows a wonderful design by an All-Knowing Creator. I mentioned in the chapter titled "God's Terrarium" a little bit about how ozone is made. If you remember, UV light from the sun breaks apart the oxygen molecule (O^2), freeing up single oxygen atoms (O^1), which will re-combine with two other oxygen molecules making O^3. This combination produces ozone that acts like an opaque filter to UV sunlight. However O^3 is an unstable element, which means that it eventually will revert back to a normal oxygen molecule (O^2).

Now if it takes sunlight to make ozone and its dark at the South Pole for six months, no ozone would be produced for that period of time. The ozone that was there (prior to the arrival of winter), re-converts back into normal oxygen during the winter darkness leaving a large hole every spring. This is a natural happening since the beginning of time. Thus, God has made sure that as long as there is sunlight and oxygen, there will always be ozone! God didn't create the earth in vain He *"formed it to be inhabited."*

So why do NASA and some scientists continue to make a big issue of this? The answer is easy, $$$$. *"What you have to understand, says NOAA's Melvyn Shapiro, is that this is about money. If there were no dollars attached to this game, you'd see it played in a very different way. It would be played on intellect and integrity."*[3]

Environmental myth #2. The Greenhouse effect.

Two years ago I was invited by the Professor of Environmental Science to give a presentation to his senior class on environmental science. He wanted me to lecture them on environmental science from a creationist point of view. I began my lecture by asking them this question: "What are the five primary environmental problems the earth faces?" These are the problems, and they rated them in order of importance:

1. Destruction of the ozone layer.
2. The greenhouse effect, which will melt the polar ice caps, causing massive flooding of the world's coastal cities
3. World's forests are being destroyed.
4. The world is over-populated.
5. Acid rain is killing our lakes and forests.

Author, Stephen Budiansky, writing for US News & World Report (December 13, 1993) writes: *"This much is certain: carbon dioxide, water vapor and several other atmospheric gases trap heat that otherwise would radiate form Earth, leaving the planet to freeze. And since 1750, the concentration of carbon dioxide in the atmosphere has increased from 275 ppm to 355 ppm as the burning of fossil fuels has expanded.*

"Computer models of the atmosphere calculate that a doubling of the carbon dioxide could occur in the next century if no action is taken to limit emissions, will cause a 1-to-5-degree Celsius rise in average global temperatures." (A 1-to-5-degree Celsius change would be equivalent to a 1.8 degree change on the Fahrenheit scale).

Most stories you are likely to hear about carbon dioxide (CO_2) are scary and present a real "doomsday" scenario to the reader. An environmentalist writes: *"We have to offer up scary scenarios, make simplified, dramatic statements, and make little mention of any doubts we may have. Each of us has to decide what the right balance is between being effective and being honest."*[4]

The truth about CO_2 is that it is as essential to life on this planet as oxygen. The question we should ask then, is there too much CO_2? An easy way to answer that question would be with a question. Do you like a world with lots of trees and vegetation? If your answer to that is yes, then bring on the CO_2! Carbon dioxide is the primary raw material used by plants to produce food by the process of photo-synthesis. Therefore, the more atmospheric carbon dioxide, the more luxuriant will be the plant growth on our planet, and the more luxuriant plant growth, the more oxygen those plants will produce. It's just one more sign of God's wonderful balance of all life forms around us. Each is dependent upon the other for survival. One couldn't have risen without the other.

Also consider these possibilities. As the atmospheric content of CO_2 rises, deserts will shrink as plants acquire the ability to survive and reproduce where before, it was impossible for them to grow. With enhanced plant growth, more organic matter will be returned to the soil producing a whole host of beneficial consequences. Earthworms, for example, would increase proportionately, since soil organic matter is what sustains them. Their increased activity in turn, will build soil structure and fertility while enhancing aeration and drainage. The world over-all, would become a more productive place.

Note: Since the original writing of this book in 1997, I just read an article in "Environment and Climate News," Dated October 2009; this enlightening article by Bonner Cohen, a research scientist with the African Research Unit titled: "Sahara Desert Greening." Mr. Bonner wrote in part:

"Satellite images of the Sahel—the vast, belt-shaped region south of the Sahara stretching from the Atlantic coast to the horn of Africa—show a steady north-ward march of grasses, shrubs, and even trees since the early 1980s."

This new data was collaborated by Stefan Kroepelin, a climate scientist at the University of Cologne in Africa. Dr. Kroepelin has done research on the Sahara now for more than 20 years. Dr. Kroepelin states: *You see birds, ostriches, gazelles coming back, even sorts of amphibians coming back. The trend has continued for more than 20 years. It is indisputable."*

As I stated more than 20 years ago, CO_2 is a highly beneficial gas and the more humans produce, the more productive the earth will become. Some scientists even refer to CO_2 as "the gas of life."

Dr. Sherwood B. Idso, a research physicist with U.S. Water Conservation Laboratory in Phoenix, Arizona found that in experiments with orange trees exposed to increased concentrations of CO_2, that the trees, after a 30-month period, grew to more than twice the size as orange trees exposed to normal atmospheric air. In fact, Dr. Edso concludes: *"It well could be that the rising CO_2 content of Earth's atmosphere is actually a blessing in disguise and one of the better things that could happen to mankind and nature."*[5]

But, you ask, how about the rising temperature melting all that ice in the polar regions? Before I answer that question let me say this. There is no evidence that the earth's average temperature has risen at all in the last one hundred years.

Remember the quote I just gave about having *"to offer up scary scenarios, make simplified, dramatic statements...."*? This statement was made by Dr. Stephen Schneider of the Colorado-based National Center for Atmospheric Research. Mr. Schneider indicted himself as being untrustworthy by also stating: *"each of us has to decide what the right balance might be between being effective and being honest."*

Well the same Stephen Schneider is designer of one of five climate models most frequently quoted by environmentalists to support their claims of a global warming trend. In a scientific paper that just crossed my desk titled: *"The Greenhouse Effect"* by Patrick Michaels, he writes: *"....of a hundred odd scientists in the world actively involved in the study of long-term climate data, only one - James Hansen of NASA - has stated publicly that there is a 'high degree of cause and effect' between current temperature and human alteration of the atmosphere."*

Land sat satellites that have been in place now for more than thirty years and are capable of detecting world temperature changes to within 100th of a degree, show an average temperature drop of about 1/10 of a degree in the last twenty years. In an article published in the American Legion Magazine (October 1993), an environmental scientist by the name of Ronald Bailey said: *"There is no upward trend in the globe's temperature. For the past 14 years it has been essentially stable. Global warming is another myth."*[6]

Oh yes, about the polar ice caps melting if the earth warms up. Let's assume that the earth warms up an average of 5^0f in the next fifty years. If you have ever done a study on the Antarctic region, you would have learned that the average temperature of the Antarctic, for example, is approximately -41 degrees Fahrenheit. If we warm that up an average of 5 degrees, or to 35^0f below zero, how much ice is going to melt? It is also a known fact, that recent measurements have shown the glaciers in Antarctica to be growing, not melting.[7]

So how much CO_2 do humans pump into the atmosphere on a daily basis? The closest estimate given so far is approximately 70 million tons a day. Sounds like a lot doesn't it? But consider this statement made by Dr. Roy Spencer, who was the U.S. science team leader for the advanced microwave scanning radiometer on NASA's Aqua Satellite, served as senior scientist for climate studies at NASA's Marshall Space Flight Center, and received The

American Meteorological Societies special award for his satellite temperature based monitoring work. Dr. Spencer states: *"The 70 million tons of CO_2 produced by humans on a daily basis is equal to 0.00000083% of the total gases making up our atmosphere and is an insignificant amount."*

Dr. Spencer made the above statement before a panel of scientists in Chicago in 2004.

Meanwhile, Dr. William Gray professor of atmospheric science and meteorology, Colorado State University puts it even more succinctly stating: *"This small warming is likely a result of the natural alterations in global ocean currents which are driven by ocean salinity variations. Ocean circulation variations are as yet little understood. Human kind has little or nothing to do with the recent temperature changes. We are not that influential."*

In fact, in 2008 there were more than 31,000 American scientists who signed a petition that was presented before President Obama. The following petition is a verbatim copy of those that signed. These are some of the most qualified atmospheric scientists in the United States and were sent to our government asking them not to sign the Kyoto protocol and similar agreements on global warming. My question then is: will our government officials listen, or will they play their political games that will ultimately destroy our economy? Here is the petition:

"We urge the United States Government to reject the global warming agreement that was written in Kyoto, Japan in December, 1997, and other similar proposals. The proposed limits on greenhouse gases would harm the environment, hinder the advance of science and technology, and damage the health and welfare of mankind.

"There is no convincing scientific evidence that human release of carbon dioxide, methane, or other greenhouse gases is causing or will, in the foreseeable future, cause catastrophic heating of the earth's atmosphere and disruption of the earth's climate. Moreover, there is substantial scientific evidence that increases in atmospheric carbon dioxide produce many beneficial effects on the natural plant and animal environments of the earth." Signed by 31,478 American Scientists

So why all the hoopla over a non-issue? It's very simple….politics and money. Lots of money. It is estimated that Al Gore has already made more than ninety million dollars selling carbon credits. Credits, that if Cap and

Trade becomes law, every company in the country will have to buy, sell or trade in order to achieve the goals set by the Federal Government.

Incidentally, Al Gore justifies his huge carbon foot print simply by telling people that he buys carbon credits. And he does indeed buy carbon credits; from the company he owns!

Environmental myth #3. The world's forests are being destroyed. This caption appeared in a "Time For Kids" Magazine: *"Cutting down forests makes global warming worse. Trees help absorb excess carbon dioxide from the atmosphere."* (Sept. 29, 1995)

Meanwhile, on the public television science program *Newton's Apple,* kids were being told how the rain forests "support life on the entire planet by producing oxygen." The rain forests were even referred to as "the lungs of the earth."

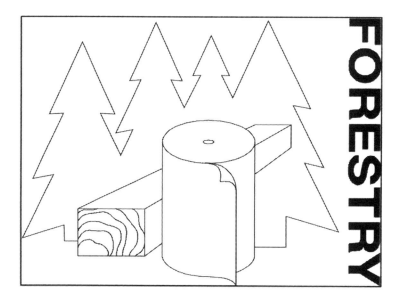

I have personally given many environmental presentations in the grade schools, high schools, colleges, and to many civic groups and organizations. I can also tell you from first hand experience that the average American, in all of these groups believes that mankind is destroying the forests of the world.

In fact, 300 Black Hills eighth grade students were asked to write an essay in 1993, on how they viewed logging in forested areas of our land. These are some excerpts from that essay, they are a real shocker: *"Thousands of trees in rain forests and normal forests are being cut down every second. I think we need to protect forests. You might want to make or join a club that helps save the trees."*

"I think the government should take away paper to save more trees...."

"In the forests when loggers cut down too many trees they make things called mudslides."

"So why don't we start to recycle instead of cutting trees?"

"Trees, America's #1 resource, are our only source of oxygen. Without them we would die."

"So if we did not have trees, we would not have oxygen and no one would be living, or a lot of people would die sooner."

"The main thing we get from trees is oxygen."

"If we did not have any trees we would not be able to breathe. But minute by minute the forest is being destroyed to build <u>your</u> homes and downtowns! The earth is overly polluted because of the factories, cars, and one of the biggest causes of all is sawmills. They are also the largest cause for disappearing trees."

"....if people keep cutting down trees we won't have any oxygen left....The sad thing is it is not just happening in the rain forest it is happening all over the world and it is happening fast!"

"I would rather be saving the forest from the tree cutters rather than seeing them being cut down. Trees are a man's best friend....Trees are just like humans, it's like people are killing trees....It's just not fair because they're killing trees for wood. It should not be happening....It should be illegal to cut down trees in the United States."

"We would die cause we can't breathe. Lots of areas are getting where there isn't any air because loggers are cutting down trees and that's providing less air....Even the wildlife needs trees. With loggers cutting down trees, we have less animals with less animals that means less people with less people who knows maybe the world will end."

Kids are literally being taught that they have no future unless they serve nature! This kind of teaching will instill fear into their young lives. They are taught this in the schools, they see it on television, and they read it in the newspapers. In fact, an article that appeared in Parade Magazine, dated March 1, 1992, and written by Dr. Carl Sagan states: *"The burning of the Brazilian rain forest. Developers eager for quick profits and poor farmers are torching Amazonia, destroying growing trees that moderate global warming. In a few decades, at present rates, there will hardly be any tropical forests left on earth."*

I can only say of Mr. Sagan, either he is writing from a spirit of paranoia, hype, or ignorance. As a professional forester of twenty-seven years, I can honestly tell you Mr. Sagan doesn't have the slightest idea of what he's talking about. Let's look at the real scientific facts.

The environmentalists claim that forty million acres of tropical rain forest are destroyed each year. Mr. Sagan states that *"in a few decades, at present rates, there will hardly be any tropical forests left on earth."*

Forty million acres a year is equivalent to a football field every second. But what is the truth? How many acres are actually being cut each year? And why is the claim made that *"forty million acres"* are being cut and burned every year?

In 1988, a Brazilian scientist was using sensors on a U.S. weather satellite to count the number of fires burning in the Amazon. He estimated the size of each fire by the area covered by smoke and guessed that 40 percent was burning in recently cleared forests and then multiplied. The number he came up with was incorporated in a widely cited 1990 report by a private World Resource Institute. Since this organization is located in Washington, D.C. it all looks pretty official. In fact, Al Gore used this report to help justify signing the Biodiversity Treaty and other initiatives to protect the global environment.

Recently, two American research scientists took a more careful look at what's really being cut and burned in the Amazon. Using more than 200 photographs of the Amazon region taken by Land sat satellites on clear days with no smoke; they compared images from the 1988 weather photographs and entered into a computer the exact measurements of each area which had been cut. They found that the average rate of rain forest loss was 3.7 million acres each year. This is a far cry from the forty million acres that is <u>still</u> being claimed.

The Amazon rain forest is the largest rain forest in the world, covering an area of 2.5 million square miles. Two and a half million square miles of rain forest equates to: one billion, six hundred million acres. If the rain forest is being cut at a rate of three million seven hundred thousand acres each year, and no re-growth occurs, it would take 432 years to cut down the entire Amazon rain forest! Numbers don't lie, people with political agendas do.

But let's take the second part of this environmental fairy tail. The one about the trees of the world producing all the oxygen. If all the trees in the world were cut down tomorrow, it wouldn't drastically affect the world's ability to produce enough oxygen, that's because two thirds of the world's oxygen comes from the ocean. All growing vegetation will produce oxygen, including plants living in the oceans of the world. There are microorganisms found in all bodies of water, the oceans and fresh water alike. These tiny organisms, called "diatoms" are so small, that it would take four to span the breadth of a human hair (about 25 microns).

While small in size, diatoms are a huge component of the earth's biota *"making up about 90% of all living organisms in the ocean."*[8] Diatoms are photosynthetic phytoplankton, and as such, produce most of the world's oxygen supply, and like a true plant, diatoms absorb most of the world's carbon dioxide load. *"As photosynthetic autotrophs, they manufacture their own nutrition, which in turn represents a substantial percentage of the earth's annual production of organic carbon (130,000 million tons). In the process they are significant consumers of carbon dioxide and are responsible for the production of much of the earth's atmospheric oxygen."*[9]

Here in the United States, I can attest to the fact that there are more trees growing in our country today, than at any time in our history. Prior to the arrival of the European settlers, fires burned on this continent every summer and there was nobody around to put them out. With the advent of the early settlers, fire was looked upon as an enemy, a force to be dealt with and so he began to wage a relentless war against this foe. As fires were contained to smaller and smaller areas of forest land, the trees grew thicker and the forests became more crowded.

By the early 1900s, our forests were so over-crowded with trees that once fire got going, nobody could stop them. Large fires began to ravage the country and in 1900 alone, great fires burned forested areas equal in size to the states of Virginia, West Virginia, Maryland, and Delaware. But as the need for wood products grew, logging and other timber management programs began to thin the forests down a little, removing some of the tremendous amounts of woody fuels that attributed to the catastrophic wild fires.

Today, contrary to what you continually hear the forest industry has done an excellent job in managing our forested lands. Rather than plundering the forests, U.S. lumber and paper industries have steadily expanded forest output during the past four decades. In 1952, for instance, there were 664 million acres of U.S forestland, containing about 610 billion cubic feet of growing stock. By 1987, there were 728 million acres of forest land with 756 billion cubic feet of growing stock, a 23.9 percent increase. By the way, it's interesting to note, that large damaging forest fires are beginning to make another come back. As we discontinue our forest management practices, the trees are again becoming thicker. With higher fuel loading values in our forests across the nation, I predict we will see a return to large, catastrophic fires, the fires of Yellowstone are a good example!

We were placed on this planet by God and told specifically by Him to manage the earth (Genesis 1:28), but we have become worshippers of the creation, rather than the Creator, and in doing so we are allowing nature to manage it for us. As a nation, we will suffer the consequences.

Environmental myth #4. <u>The world is overpopulated!</u>

An article that recently appeared in the Rapid City Journal and written by Bob Wiemer was titled: *"Explosive population growth is the root of many evils."*

Abortion is, of course, one of the many evils that Wiemer is referring to. Knowing that this country kills almost two million unborn children every year, I launched into a study concerning world population.

I remember reading a little from Paul Ehrlich's book titled: *"The Population Bomb"*. In this book of paranoia and hype, Mr. Ehrlich, who is a Professor at Stanford University, wrote: *"In the 70s the world will undergo famines - hundreds of millions of people are going to starve to death."* Mr. Ehrlich also refers to pregnancy as *"a nine month disease."*

During this same time, Nobel laureates were telling Congress that unless population growth stopped, a new dark age would cloud the world and *"men will have to kill and eat one another."* Another well regarded book *"Famine 1975"* predicted that hunger would begin to wipe out the Third World that year! Of course we now know that none of these dire predictions of paranoia and hype came to pass.

Consider for a moment the lament of this environmentalist: *"Plowed fields have replaced forests; domesticated animals have dispersed wild life. Beaches are plowed, mountains smoothed and swamps drained. There are as many cities as, in former years, there were dwellings....Everywhere communities, everywhere life....Proof of this crowding is the density of human beings. We weigh upon the world; its resources hardly suffice to support us....In truth, plague, famine, wars, and earthquakes must be regarded as a blessing to civilization, since they prune away the luxuriant growth of the human race."*

Sounds familiar doesn't it? This statement was made by Tertullian in the year AD. 200! As a matter-of-fact, our American ancestors also complained that the land was filled up even when their largest cities only had 30,000 people in them.

In 1972 a group of researchers at MIT would issue *"The Limits of Growth"* which used advanced computer models to project that the world would run out of gold in 1981, oil in 1992, and arable land in 2000. Civilization itself would collapse by 2070. These same educated people were predicting an ice age by the end of the 80s. Now, of course, they're prognosticating just the reverse; the earth is heating up because of the green house effect. As a mater-of-fact, in the Rapid City Journal article I quoted at the beginning of this discussion, Wiemer's wrote: *"The deserts are expanding; the oceans are rising; the air is growing foul; scarcity evokes tribal rivalries, and forests the size of Vermont, New Hampshire and Massachusetts go under the ax every year."*

So you might be asking yourself, what are we supposed to believe? First of all, let's start with the truth. Let's forget about hype and paranoia and stick with numbers. Numbers don't lie, as I mentioned before, people and governments with political agendas do.

One reason people and governments use the "over-population" issue is to justify abortion in the eyes of the general public. It is far easier to soothe one's conscience after killing an unborn child if he or she believes the world is over-populated.

Is the world over-populated? Let's take a look at this issue from a logical stand-point, using real numbers.

Let's assume that every person in the world was brought together into one place. We would stand all these people shoulder to shoulder, front to back. Assuming that each person would need a space of ten square feet to stand in, how large an area would be required to hold them?

Would the entire United States be required to hold them, or would they all fit in an area the size of the State of Alaska? Actually the State of Alaska would be much too large, so would the State of Texas. Ok, how about Rhode Island which is the smallest State in the Union, would that hold all the people in the world?

Now I'm not saying all the people of the world, if brought together could live there, I'm giving an example of how much of the world's surface would be covered by people.

Today there are about 6 billion people in the world. If each person were assigned a ten square ft. area to stand in, then we can easily see that an area of 60 billion sq/ft. would be required for every person in the world.

I live in Pennington Co., in the state of South Dakota. Pennington County is 2776 sq/miles in area. There is a total of a little more than 27 million sq/ft in a mile. Thus, 27 million X 2776 = a little more than 75 billion sq/ft of area in my county. But, as you just saw, only 60 billion sq/ft. of area is needed. Pennington County is too large!

Again, using numbers that don't lie, we arrive at a simple fact: all the people in the world would fit in a one thousand square mile area. This means if you got into your car and drove north for 32.6 miles, then east, then south, then west again, you would have covered an area that all the people of the world could fit inside of!

The overpopulation myth ushers in this quote that environmentalists are fond of using: *"Twenty percent of the world's population uses eighty percent of the world's resources."* I would like to submit to you the truth. Eighty percent of the world's resources are produced by twenty percent of the world's population.

The environmentalist declares: *People are primarily parasitic, consuming the bounty of the world."* Again the truth: people produce the bounty of the world; for every mouth that is born into it, comes a pair of hands to work and a brain to create. Some of the most densely populated places in the world are also the wealthiest places in the world, having the highest standard of living; take Tokyo, London, New York or Paris for examples.

The environmentalist will tell you: *"Worldwide famine is increasing, as food becomes more scarce."* Truth: food production in the developing world more than doubled between 1965 and 1990, according to a recent article by population scholar John Bongaarts in the *Scientific American.* Calorie intake per person in developing countries increased an extraordinary 21 percent. The number of chronically malnourished people fell 7 percent in the 11 years from 1979 to 1990. Lester Brown, president of the World Watch Institute stated: *"Much of the world is better fed today than it was in 1950."*

And part of this is because world food crops are increasing every year; in part because of better farming practices, but also from the small increases we see in the atmosphere from that polluting gas....CO_2.

According to scientists with the Food and Agricultural Organization of the UN:

"By using modern agricultural methods the Third World could support more than 30 billion people."

Environmental myth #5. <u>Acid rain is poisoning our lakes and killing our forests!</u>

This myth has been around as long as the ozone hole. Back in the late 1970s environmentalists began to point out air pollution coming from power generating plants in the eastern half of the US. Coal fired power plants produce large amounts of sulfur dioxide (SO_2) and nitrous oxide (NOx).

The theory is that these gasses react in the atmosphere with water, oxygen, and other chemicals to form various acidic compounds resulting in a mild solution of sulfuric acid and nitric acid. Winds aloft then would carry these compounds over long distances where they are deposited into the various lakes and forests by the rain, which they quickly dubbed "acid rain."

While I have no doubt these chemicals were occurring in the atmosphere and, as I stated, forming a mild solution of sulfuric acid and nitric acid which was falling on forests and lakes; my contention is, that like everything most environmentalists do, it is blown clear out of proportion in order to frighten people and get them up in arms with the intent of eventually either shutting down the power plant or making it very costly to operate.

One might ask the question why they (environmentalists) would want to do this. David Foreman in his book "Ecodefense" gives some clues. He wrote: *"We must make this an insecure and uninhabitable place for capitalists and their projects. This is the best contribution we can make towards protecting the earth and struggling for a liberating society."* Vol. 10

Carl Amery, quoted in the book "Trashing the Planet:" *"We, in the green movement, aspire to a cultural model in which the killing of a forest will be considered more contemptible and more criminal than the sale of 6-year-old children to Asian brothels."* Nice guy huh?

These people are radical leftists and worship nature as their god. I could go on with hundreds of quotes from the environmentalists and their hatred for America and her capitalist form of government. But, they also have a powerful ally; Hollywood and the news media. And today, we can see how they are making headway in destroying industry in our country and have been literally responsible for putting millions of Americans out of work.

We live in an extremely resilient world, but we also need to be good stewards of this world the good Lord has placed us in. I don't believe that factories should pollute the air, and since the rise of radical environmentalism, the factories have had to clean up their act—and that's a good thing. But, their goal isn't to protect the environment, but rather to destroy our American way of life.

Environmentalists claim that over 16,000 lakes have been killed by acid rain. So let's check it out. Does acid rain kill lakes and forests?

Again, I'm not trying to leave you with the impression that acid rain is good. That's not at all what I'm saying. Power companies need to scrub their coal-fired smoke stacks as well as they can, after all that's part of being a good steward and good stewards are good environmentalists; which is what we all need to be, not radical leftists using the environment to bring about our purpose.

Before I retired from the South Dakota Division of Forestry, I had access to the Game Fish and Parks Dept. biology lab which was next door to our offices. Several of the fisheries biologists working there were good friends of mine and over the years we drank a lot of coffee together.

On one such coffee day the subject of acid rain came up and I could tell by their reaction that they weren't too impressed by the problem. So I asked them if this was just another non-issue like so many of the scenarios that radical environmentalists were pushing. They quickly responded that it definitely was a non-issue.

I further queried them. "So tell me" I said, "what are the facts on this?"

"Well," Ron said: "the biggest problem with the lakes in the Northeast part of the country is they are bedded in granite basins."

"So what has that got to do with anything?" I asked. "Everything" Ron replied. "As you well know, the forests making up that part of the country are Boreal, and boreal forests contain large numbers of tannin oak trees."

I was beginning to get the picture. Tannin oak is called this because the early settlers found the leaves and bark contained large amounts of algal acid. This was used by them to tan wild animal skins.

"So the algal acid leaches into the lakes making them to acidic for fish and other forms of aquatic life. "But what about the granite basins these lakes are located in, what has that got to do with it? After all, many lakes are surrounded by forests with tannin oak and other acid producing trees and they are unaffected."

Ron responded, "That's exactly right, but those lakes are bedded in limestone basins. Limestone acts like an antacid you take for acid indigestion. It buffers out the acids! Ron continued, "Don those lakes in the N/E have probably always been dead because they are bedded in Granite and not limestone! Further, if you study any history written by those early settlers they knew which lakes you couldn't fish in because there were no fish."

And as Paul Harvey would say: "Now you know the rest of the story." We have spent hundreds of millions of dollars studying another non-issue that made major news because of radical environmentalism.

And as a forester I can tell you that acid rain falling on forested areas have minimal to no effect on the trees, otherwise they would have all died out during the industrial revolution when the air was so polluted with coal smoke it darkened and stained the bark on all the trees growing in those areas.

I am amazed at the influence Hollywood has on the American people. They produce a movie that is cleverly disguised as entertainment but in reality is sending a political message. Jane Fonda's 1970s movie, "China syndrome" comes to mind.

This movie was about the melt-down of a nuclear power plant and it depicted the reactor core melting clear through the earth; totally absurd, right? But it sent a clear message and since the filming of that movie, not one new nuclear power plant has been built in this country. Albeit, the incident at Two Mile Island, helped make Ms Fonda's movie more credible.

Even magazine ads and web sites with pictures can say a lot, even though they don't make much sense. Take the picture below that was used on a Sierra Club web site.

The caption read: *"The tree on the left is being exposed to Acid Rain, though it hasn't taken its effect the leaves are facing downwards and the tree looks weak."* The first thing one should ask himself is, "why only one tree and not the one standing beside it? Wouldn't they both be affected equally?"

And of course one can go to any forested area and find trees that are sickly because of a host of differing things.

References

1. Robert W. Lee, "The New American" p.12 June, 1992
2. Micah Morrison, "The Wizards of Ozone" Insight Mag. April, 1992
3. Op. Cit.,
4. S. Schneider, "Our Fragile Earth" Discover Mag. P.47 Oct. 1989
5. Idso, "Carbon Dioxide Can Revitalize the Planet" OPEC Bulletin, March 1992 p.27
6. Ronald Bailey, "The Eco-Crises Myth" American Legion Mag. P36
7. R. Sedjo & Marion Clausen, "Prices, Trade, and Forest Management" Econ. Update, Reason Foundation, Vol. 3 1989
8. M. Armitage and R. Lumsden, "Microgeometric Design of Diatoms-Jewels of the Sea" Acts & Facts Impacts Bulletin #265 July, 1995
9. Op. Cit.,

Notes

GOD'S LAWS

Did you ever wonder why the children of Israel could be led out of Egypt, see the great miracles of God, then turn around in a few short months and worship the golden calf? (Exodus 32:4)

I have been totally amazed that any people could be so callous and arrogant, yet it dawned on me one day that we are exactly like they were. We cannot see God face to face, so we cast off His provision and call it "luck." Let me explain.

I remember hearing a story (fable) one time about a farmer who was on-top of his barn working on the roof. He stumbled and fell from the highest point of the roof and just as he was falling he called out to God: *"God help me!"* About half way down, there was a large nail sticking out from the wall of the barn, it caught the farmer by his britches and held him firm stopping his fall, to which the farmer replies: *"That's okay God, this nail caught me."*

The farmer didn't see God's hand dart out and grab him. He didn't see an angel fly up, catch-a-hold of him and gently let him down. He didn't recognize God's provision for his life because all he could see was the nail! A physical entity.

In this chapter, I will discuss the physical laws that God has established that govern, not only the universe, but to a large degree, our lives. While these laws are invisible, God uses them to bring about His purpose for the universe and, indeed, our lives.

In Psalm 119:90,91 we find these verses: **"Your faithfulness endures to all generations; You established the earth, and it abides. They continue this day according to Your ordinances, for all are Your servants."**

You see the physical laws (ordinances) that God established to govern the universe are also His *"servants."* And He uses these servants all the time to bring about His purpose for everything that exists.

I am not trying to explain the miraculous while dealing with the physical, but rather to show that the physical was used to bring about the miraculous. For as God asked Job: *"Can you send out lightnings, that they may go, and say to you, 'Here we are!'"?* (Job 38:35) While the lightning is a physical entity, who but God could command it to be a servant; to use it as a knife if He wished?

Turning back to Exodus 14:21, we can see that God caused *"a strong east wind"* to divide the Red Sea. God could have commanded the water to part, period! But He didn't. He used a physical entity (the wind) to part the water. Further, this apparently was a cold wind because it froze the water in place, this is evidenced by Exodus 15:8 that says: *"And with the blast of Your nostrils the waters were gathered together; the floods stood upright like a heap; and the depths congealed in the heart of the sea."* The word *"congealed"* in Hebrew is *"gapha"* and means to harden, by freezing. Thus the cold wind could have been perceived to be a natural event.

We find more evidence that this is exactly what happened as we read from Exodus 14:27, *"And Moses stretched out his hand over the sea; and when the morning appeared, the sea returned to its full depth..."* Yes, when the sun came up in the morning, the ice melted and the sea returned to its normal place! Another seemingly natural event!

There is a tremendous metaphor here. For the children of Israel departed Egypt in the night (Exodus 12:14) and finished coming up out of the Red Sea on the third morning. They were delivered at that time from bondage as their captors were buried in the depths of the sea.

Jesus Christ, the Son of God was raised from the grave on the morning of the third day. He likened our sins to a mountain that we are in bondage to, and He said of that mountain: *"....also if you say to the mountain, 'Be removed and be cast into the sea, it will be done."* (Matthew 21:21) Egypt itself, is a symbol of sin (Revelation 11:8) and Egypt was Israel's mountain, God was Israel's provision but they didn't recognize it.

Likewise, Jesus Christ is our provision for freedom from bondage and He, and only He, can bury our sins in the deepest ocean.

The Bible is replete with examples of how God used the forces of nature to bring about His divine will. When the children of Israel crossed the Jordan River, for example, we are told they crossed over on dry ground: ***"Then the priests who bore the ark of the covenant of the Lord stood firm on dry ground in the midst of the Jordan; and all Israel crossed over on dry ground, until all the people had crossed completely over the Jordan."*** (Joshua 3:17)

The reason the Jordan stopped flowing as the children were about to cross is because a landslide 15 miles upstream dammed up the river! ***"....and as those who bore the ark came to the Jordan, and the feet of the priests who bore the ark dipped in the edge of the water (for the Jordan overflows all its banks during the whole time of harvest), that the waters which came down from upstream stood still, and rose in a heap very far away at Adam, the city that is beside Zaretan. So the waters that went down into the Sea of Arabah, the Salt Sea, failed, and were cut off; and the people crossed over opposite Jericho."*** (Joshua 3:15,16)

Halley's Bible Handbook says: *"At Adam, where the waters were held, the Jordan flows thru clay banks 40 ft. high, which are subject to landslides. In 1927 an earthquake caused these banks to collapse, and so dammed up the river that no water flowed past them for 21 hours."*[1]

The point to all this is that God directs natural laws to fulfill His divine will, and these laws, are His servants. Let's take a look at some of these laws and the physical science that upholds them, and you will agree, Scripture deals with a science far in advance of anything we yet know or can understand.

I mentioned earlier Job 38:35 where God asked Job if he could command the lightning. Job's answer to that obviously was no, but then, neither can we. Today however, we know quite a bit about lightning and the important role that it plays on our planet. Ironically, what we are just learning through modern technology, Job had written down 3000 years ago!

In Job 28:24-26 we read this about lightning: ***"For He looks to the ends of the earth, and sees under the whole heavens, to establish a weight for the wind, and mete out the waters by measure. When He made a law for the rain, and a path for the thunderbolt."*** Again in Job 38:25, he refers to a pathway for the lightning. Does lightning follow some kind of a path?

Recently a research paper on plasma gas crossed my desk; it did a little more than raise my eyebrows and grab my attention. It was titled: *"Not With A Bang"* and was written by Anthony Peratt, a physicist at Los Alamos National Laboratory in New Mexico. Dr. Peratt wrote:

"Because of its free electrons, a plasma is a particularly good conductor of electricity. In fact, one of the most dramatic manifestations of a plasma is lightning. As a thunderstorm gathers, regions of negative electric charge form along the bases of clouds, which in turn causes positive charges to build up along the ground. The electric field between clouds and ground becomes so strong that the air is ionized. A <u>conducting path of ions</u> and free electrons - plasma - is created, through which the lightning is discharged."

The original discovery of this "pathway" was made back in 1985 using high speed cameras and film (1/10,000 sec.), to catch lightning strikes. The pictures clearly revealed a pathway of conducting gas (plasma) preceding the discharge. If Scripture wasn't inspired by God, it is inconceivable that Job could have possibly known about this 3,000 years ago.

Perhaps, even more amazing than the revelation of a pathway for lightning is found in the next verses of Job: **"Listen to this, O Job; stand still and consider the wondrous works of God....Do you know the <u>balance of clouds</u>, those wondrous works of Him who is perfect in knowledge?"** (Job 37:14 & 16)

This article was written by meteorologist David Ausman who does meteorological research work with the Tiros weather satellite data: *"High altitude airplanes recently discovered that there was a high electrical energy flow upward from the top of thunderstorms to the ionosphere. This showed a thundercloud may be considered as a negative electrical pump conveying a negative charge to the earth and a positive charge to the upper atmosphere. Through investigation it was learned that the ionosphere continually discharged a current of 1,800 amperes to the earth over the whole globe. Instruments placed on research missiles reveal the total charge of the ionosphere would discharge completely within 5 minutes creating an electrical imbalance. Thunderstorms supply the reverse current necessary to maintain potential between the earth and upper atmosphere. It was estimated to take 1,800 thunderstorms at any one instant to <u>maintain this electrical balance</u>. The Tiros weather satellite indicates there are in fact approximately 1,800 thunderstorms occurring worldwide at any given time."*

WOW, I am humbled by the tremendous knowledge contained in this ancient text. The clouds maintain an electrical balance! Again, how could Job have ever thought up something like this? It's quite apparent there is absolutely no way he could have. God had to have told him. Oh yes, 1,800 thunderstorms going on in the world at any given time! What does my ancient text say about that: **"He (God) sends forth under the whole heaven, His <u>lightning to the ends</u> of the earth."** (Job 37:3)

Let us journey back for a moment in time. Back past the day of Job, back to a time shortly after the flood of Noah. The ark has just come to rest on *'the mountains of Ararat."* And God says to Noah: **"*I set my rainbow in the cloud, and it shall be for a sign of the covenant between Me and the earth. It shall be, when I bring a cloud over the earth, that the rainbow shall be seen in the cloud...."*** (Genesis 9:12)

Most people are aware of the significance of the rainbow and its connection to the Genesis story. What most people aren't aware of is how a rainbow achieves its most vivid state of beauty and color. The two primary ingredients of course, are the falling rain and sunshine. As the sunshine strikes the falling rain, the light is parted or refracted into the seven colors of the light spectrum. In Job 38:4, God asked Job this question: **"*By what way is the light parted?"*** Of course Job couldn't answer the question, and today, we don't know a whole lot more than Job did. We understand that bending light produces all the colors of the spectrum. We also know that when light is stretched it changes color. The degree of stretching produces the various colors.

The Genesis account of the flood tells us that it rained on the earth for forty days and forty nights and that after it quit, God produced the rainbow as a sign to all mankind that He would never again flood the earth. What most people don't realize is that the sun must strike the falling rain at an angle of $40°$ in order to produce the most visible rainbow! Just one more sign given to man from an Omnipotent Creator.

Solomon also wrote of the earth's hydrologic cycle in Ecclesiastes 1:6,7: ***"The wind goes toward the south, and turns around to the north; the wind whirls about continually, and comes again on its circuit.***

"All the rivers run into the sea, yet the sea is not full; to the place from which the rivers come, there they return again."

One only need look at a satellite picture of the earth and its weather systems to realize that the winds do rotate in huge circular patterns. Winds associated with low pressure systems rotate in a counter clockwise pattern, while high pressure systems rotate clockwise. Did Solomon have the use of satellite photos to make this determination? Of course not, he had something better...

the voice of the Spirit of God. Notice also, the statement: *"All the rivers run into the sea, yet the sea is not full...."* One way or other, all the rivers in the world end up emptying into the oceans, yet the oceans never fill up. We know this is because the amount of water evaporated from off the oceans is equal to the total amount of water running into them. The prophet Amos, also alludes to this same thing, while speaking of God, he writes: *"He calls for the waters of the sea and pours them out on the face of the earth; The Lord is His name."* (Amos 5:8)

In Proverbs 8:26, Solomon wrote: *"While as yet He had not made the earth, nor the fields, nor the highest part of the dust of the world."*

Every twenty-four hours 26,000,000 meteorites and asteroids are turned into dust because the earth's atmosphere is the correct density. These meteorites are burned up by the friction produced upon entry into the atmosphere, thus producing the dust. It has been estimated that this dust literally rains down on the surface of our planet at the rate of 14,000,000 tons each year.

Not only is *"the highest part of the dust of the world"* raining down on the world, but if it weren't for earth's protective blanket of atmospheric gasses, all terrestrial life would have been pulverized long ago. If you don't believe that, take a look at the other planets in our solar system. They have no atmosphere, at least from an oxygen standpoint, and they all look as though they've been pulverized over every inch of their surface.

Recently astronomers are wondering aloud if the other planets in our solar system aren't in there respective orbits to act as a type of vacuum cleaner for the earth. Because they seem to take all the big hits that our atmosphere couldn't possibly burn up before they impacted the earth.

These planets seem to always be in the right place at the right time to draw comets, large meteors and other space debris onto their surface through gravitational attraction. This speculation rose in 1994 when the comet Shoemaker-Levy 9 slammed into Jupiter. Beginning on July 20 1994, a string of comet pieces, from the breakup of comet Shoemaker-Levy 9, bombarded Jupiter for nearly a week.

Then again, on July 20, 2009 another comet or large meteor impacted the Jovian surface. Coincidently, July 20, 2009 was the 40th anniversary of man's first landing on the moon.

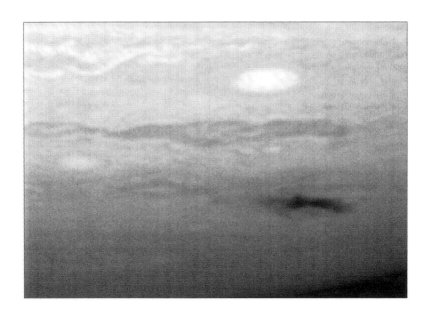

"The black mark on the picture above shows where the object impacted the Jovian surface." Hubble photo.

Consider for a moment, had it not been for Jupiter, the earth could have taken this hit. And since Jupiter is thousands of times larger than the earth, and looking at the size of the impact point, it looks like that would have been the end to all life on our planet and quite possible, the end of our planet period.

Also, the Moon, Mercury and Mars are all pock-marked with huge impact craters. I'd say someone is definitely watching out for our little blue sphere we call home.

In 1976, the research submarine "Alvin" photographed fresh-water springs in the depths of the ocean. Our knowledge of the existence of submarine springs has been only recent. While these marine biologists and oceanographers were the first to discover the springs, Job, had actually recorded their existence 3,000 years ago, and he did it without the use of a submarine. In Job 38:16, God asked him these questions: ***"Have you entered the springs of the sea? Or have you walked in search of the depths?"***

The story of Matthew Fontaine Maury is another amazing example of divine inspiration of the Scriptures. *"Pathfinder of the seas"* is the name given to Matthew Maury because he was the first person to chart and map the ocean currents. His son was reading the Bible to Maury while he was recovering from an illness and happened to read Psalm 8:8 which says: *"The birds of the air, and the fish of the sea that pass through the paths of the seas."*

This verse gave Maury the idea of circulating ocean currents and he promptly went out and discovered the truths of scripture. A statue to the memory of this great scientist stands in Richmond, Virginia.

And of course, we shouldn't forget Christopher Columbus, who, upon convincing Ferdinand and Isabella, King and Queen of Spain, to finance his trip to what he believed would be the discovery of a route through the West Indies.

Columbus believed the world to be a sphere by virtue of the Holy Scripture. This wasn't a popular belief to be sure, for the science of the day taught that the earth was flat. How many times did the people mock him with words like: *"Ah, here comes our vagabond wool carder again, with his pathetic prattling about spheres and parallels. Tell us, Cristoforo, does the world appear any rounder to you today?"*[2]

Christopher however, stood firm in his belief. He was well aware of what Isaiah, God's great prophet, wrote, and its implications: *"It is He who sits above the circle of the earth...."* (Isaiah 40:22) Columbus was also sure of his mother's prophecy over him when he was yet a child. It was her contention that God had given her a verse concerning her young son. This verse also was found in Isaiah and she never let him forget it. The verse is prophetical in nature and I believe was divinely given by God to the young boy's mother concerning her son: *"I will give you as a light to the nations that my salvation may reach to the end of the earth."* (Isaiah 49:1-6) It is also interesting to note that the name given to Columbus, "Christopher" means: *"Christ-bearer."*

There is no doubt about it, America was established by God Almighty to be a *"light to the nations"* and she fulfilled this purpose as no other nation has and became the most blessed nation on the face of the earth. Sadly enough, America seems on the verge of rejecting the God that made her great, if this happens, I fear for her future.

Recently I had the opportunity to debate an atheist whose contention it was, that the Bible does not teach that the earth is a sphere, but rather to be flat. "Oh yes" said he, "it teaches that the earth is a round circle, but then a dime is a round circle, and it is also very flat."

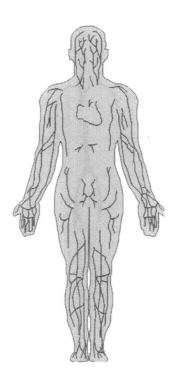

"A good argument" I asserted, "but you are forgetting what Job wrote a thousand years before Isaiah came along." I picked up my Bible and quoted Job 26:10: ***"He drew a circular horizon on the face of the waters, at the boundary of light and darkness."***

"A circular boundary on the face of the earth could only happen if the earth was a sphere. Otherwise, the entire face of a flat surface would be lit by the sun and there would be no circular boundary." He quickly moved on to another subject.

The dust of the earth!

If we were to do a chemical analysis of dust, we would find this composition:

- Hydrogen 60%
- Oxygen 25.5%
- Nitrogen 2.4%
- Carbon 10.5%
- Calcium .22%

- Phosphorus .13%
- Potassium .04%
- Sulfur .13%
- Chlorine .03%
- Sodium .03%

If we were to do a chemical analysis of the human body, we would find the exact same elements. Is it a mere coincidence then that the Bible says as God made man: **"And the Lord God formed man of the dust of the ground, and breathed** *into his nostrils the breath of life."* (Genesis 2:7)

The Psalmist wrote, concerning the human body: *"I will praise you, for I am fearfully and wonderfully made."* (Psalm 139:14)

How accurate is this statement? And just how complex is the human body? Let's make a quick examination of the body and see if we have been *"fearfully and wonderfully made."*

Starting with the DNA (Deoxyribonucleic acid), your body is made up of approximately 30 trillion cells.

DNA directs each cell to become what you are. DNA tells your legs exactly how long each needs to be, your arms and each finger. It decides the color of your eyes and your hair. It is a super coded blueprint for your entire body. Dr. Kent Fish, who prior to his death was Professor of Biology at the South Dakota School of Mines and Technology, told me in a personal conversation that the idea that DNA could have come about through random processes *"was totally absurd and based on man's ignorance."*

It is estimated that within each DNA cell, there are several hundred thousand genes. Dr. Duane Gish said of these genes: *"We know very little about them, except they are very complex."*

Just how complex is the DNA. Let's check it out. Each cell has within it enough coded information to fill 2,000 books!!! That alone is mind boggling. How many books of coded information then are in your body? You would have to multiply 2000 X 30 trillion. Kind of makes the most complex computer look like a tinker toy for the simple doesn't it?

Pictured here a DNA cell looks kind of like a coiled car spring. If we took just one DNA and stretched it out, it would be over six feet long. If all the DNA in your body was placed end-to-end, it would stretch from here to the moon over 100,000 times. If all this very densely coded information were placed in typewritten form, it would completely fill ten Grand Canyons. Yet all the DNA in your body would not fill two teaspoons! This kind of complexity is very hard to wrap your mind around isn't it? How could 2,000 books of information be encoded into a cell so tiny it cannot be seen except with an electron microscope? And, and arrogant man thinks this was brought about by natural processes of mutating genes and natural selection. Then he proudly calls it "science!"

This also substantiates what John wrote about the Lord: ***"And there are also many other things that Jesus did, which if they were written one by one, I suppose that even the world itself could not contain the books that would be written."*** (John 21:25)

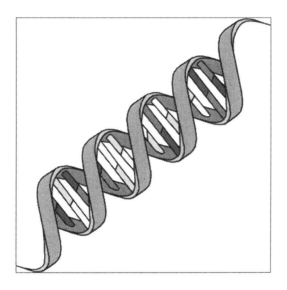

DNA Molecule

The human brain weighs approximately three pounds. It has twelve billion cells, each connected to ten thousand other cells, and thus has 120 trillion electrical connections! The brain has been referred to by scientists as the *"most powerful and mysterious entity in the universe."* This little three pound mass processes, on an average, *"100 million separate messages every second, most coming from the senses."*[3]

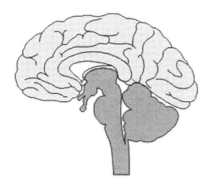
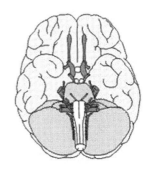

So how long, at this incredible rate of information flow, would it take to fill the brain to its capacity? According to the Moody Institute of Science, if we learned something new every second, it would take 3 million years of storage to fill it to capacity. How much storage capacity does your computer have? All the computers in the world aren't even in the running. Yet who among us would be so ignorant that he could actually believe that computers randomly arranged themselves in a complex order, and then, by the same processes were programmed?

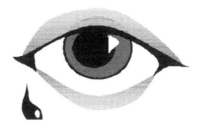
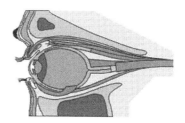

The human eye: *"While today's digital hardware is extremely impressive, it is clear that the human retina's real-time performance goes unchallenged. Actually, to simulate ten milliseconds of the complete processing of even a single nerve cell from the retina would require the solution of about 500 simultaneous nonlinear differential equations 100 times and would take at least several minutes of processing time on a Cray supercomputer. Keeping in mind that there are 10 million or more such cells interacting with each other in complex ways, it would take a minimum of 100 years of Cray time to simulate what takes place in your eye every second."*[4]

Darwin himself exclaimed: *"The thought of how the eye was formed through the processes of natural selection, makes me ill."*

The unborn child

Each of us begins life as a single cell that results from the fusion of a sperm and an egg. The cell multiplies to form a mass of cells which are all alike. How beautifully this matches the Scriptural text found in Psalms: ***"My substance was not hid from Thee, when I was made in secret, and curiously wrought in the lowest parts of the earth, Thine eyes did see my substance, being yet unperfected; And in thy book all my members were written....when as yet there were none of them."*** (Psalm 139:15,16 KJV)

Man in his wisdom, for centuries believed in the "pre-formation" theory. In this idea, the child was already pre-formed and completely developed inside the male sperm. This view was disproved when the modern microscope came into use during the late eighteenth century. Through careful scientific observation it was discovered that the male sperm united with the egg and begins to multiply in mass.

In Psalm 139:15, the word ***"substance"*** is used. The Hebrew word for substance is *"ostem"* which actually means "body," thus a better translation for this verse would read: "....my body was not hid from Thee when I was made in secret."

Then in verse 16, the word ***"substance"*** is used again. This time the Hebrew is "golem" which means unformed mass. This happens to be the present day biological definition of an embryo! How could the Psalmist have known about this fact of the early development of the human embryo when science of that day taught a completely different view? According to a scientist by the name of James Hall, when the Hebrew words are analyzed for verse sixteen, it would read: *"Thine eyes did see my unformed mass yet being incomplete and in Thy book all my parts were written, which in the required time were molded into shape when as yet there was none of them."*

The inference from these Scriptures is that conception initiates life and that somehow God is involved in all conceptions.

Scripture is certainly very plain that a child comes into existence at the very moment of conception. Consider what Job wrote lamenting the day of his conception because of his great torment: ***"May the day perish on which I was born, and the night in which it was said 'A male child is conceived."*** (Job 3:3) From the moment of conception it is a ***"child."*** And like the Psalmist, Job declares that it is God who fashions the unborn in its mother's womb: ***"Did***

He who made me in the womb make them? Did not the same One fashion us in the womb? (Job 31:15)

In our haughtiness, we naively believe that it is we ourselves who make babies and not the God of Creation. Listen to what the Psalmist says in Psalm 100:3, **"Know that the Lord, He is God; It is He who has made us, and not we ourselves...."**

Consider then, the guilt of those who would destroy the unborn as it yet lies in the secret place of its mother's womb. God's great prophet Jeremiah wrote: **"Also on your skirts is found the blood of the lives of the poor innocents. I have not found it by secret search, but plainly on all these things."** (Jeremiah 2:34)` We murder *"the poor innocents"* in the *"secret"* place, and say, we are innocent, for we did not know! But I tell you, their Creator knows, and He will be the final judge.

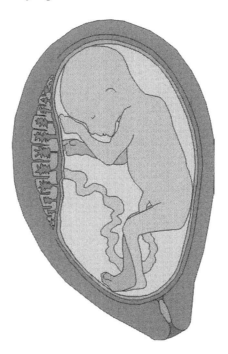

Have you ever heard of the protein molecule called laminin? Laminin is a protein that is part of the extracellular matrix in humans and animals. The extracellular matrix (ECM) lies outside of cells and provides support and attachment for cells inside organs (along with many other functions). Laminin has "arms" that associate with other laminin molecules to form sheets and bind to cells. Laminin and other ECM proteins essentially "glue" the cells (such as those lining the stomach and intestines) to a foundation of connective tissue. This keeps the cells in place and allows them to function properly. The structure of laminin is very important for its function (as is true for all proteins). One type of congenital muscular dystrophy results from defects in laminin. A mutation!

In fact, laminin is often referred to as the glue that binds our bodies together. Quite literally, it's what holds us together. You can buy T-shirts at Christian stores that have a picture of a laminin molecule on it that says; *"Great designers always leave their mark"*

This statement is obviously referring to Romans 1:19 & 20, **"What may be known of God is revealed in them, for God has shown it to them. For since the creation of the world His invisible attributes are clearly seen…."**

So what invisible attributes is the Bible talking about in this instance? Well it has just been discovered that the laminin molecule is in the shape of a cross.

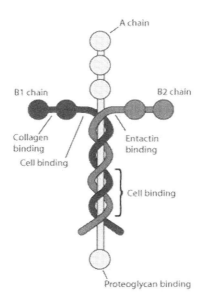

And, lest we forget the verse already quoted in an earlier chapter of this book, Colossians 1:17, **"And He is before all things, and in Him all things hold together."**

Wow, are you humbled yet? Remember, He told us 2000 years ago: ***His invisible attributes are clearly seen."***

156   Modern Science and an ANCIENT Text

So, you ask yourself, does this mean that atheists will come to a belief in a supernatural Creator. The answer to that question is, not necessarily. Consider the statement made by Richard Dawkins in his book "The Blind Watchmaker".

Dawkins wrote: *"Biology is the study of complicated things that give the appearance of having been designed for a purpose."*

Dawkins might as well have gone on by saying: "yes Lord, I see what You have shown me, but I reject it, my ideas are higher than Yours!"

So, Scripture tells us that by rejecting the truth of God *(for God has shown it to them)* they become fools. And dear reader, it is Dawkins own words that indict him, for he freely admits that *"complicated things give the appearance of having been designed for a purpose."*

However, to be fair, not all atheists are so arrogant they will continue to deny what is obvious to them. One such person, Professor Antony Flew, who, it is claimed is the world's best known atheists and noted debater of prominent creation scientists cited great advancements in science by creationists for his new belief in God as the Creator.

While doing an interview on the Hannity and Combs show on October 2007, Professor Flew said this: *"There were two factors in particular that were decisive. "One was my growing empathy with the insight of Einstein and other noted scientists that there had to be Intelligence behind the integrated complexity of the physical Universe. The second was my own insight that the integrated complexity of life itself – which is far more complex than the physical Universe – can only be explained in terms of an Intelligent Source. I had to go where the evidence led me."*

What evidence is the Professor referring to here? I would like to say it was the evidence of real science, the evidence, if followed without bias can only lead us back to the God of Creation, because **"His invisible attributes are clearly seen."**

I want to take you for a moment down to the basic level where life begins; to the bacterial flagellum motor. It is here we see complexity so awesome that it is impossible to even contemplate its origin through natural evolution guided processes.

"The bacterial flagellum is composed of about 25 different proteins, each of them in multiple copies ranging from a few to tens of thousands. It is driven

by a rotary motor with a diameter of only 30 to 40 nm. Much like the modern rotary engine, the bacterial flagellum consists of many different parts that carry out different functions, such as a rotary motor, bushing, drive shaft, rotation-switch regulator, universal joint, helical propeller, and rotary promoter for self-assembly. The filament of the bacterial flagellum is a hollow tube composed of protein, and is only 20 nanometers thick. It is helical, and has a sharp bend just outside the outer membrane called the "hook" which allows the helix to point directly away from the cell. A shaft runs between the hook and the basal structure, passing through protein rings in the cell's membranes that act as bearings."

This statement is the biological description of just one tiny part of a living cell.

The flagellum motor could not have evolved because if even one component is missing it cannot function. Kind of like a car engine. If the crankshaft didn't come with the engine it would never run. So the question is: how could it over millennia, put itself together? Now you see just another reason why Dr. Wald made the statement he did: "

"One has only to contemplate the magnitude of this task to concede that the spontaneous generation of a living organism is impossible. Yet here we are as a result. I choose to believe in spontaneous generation."

Sadly enough, Dr. Wald has looked the evidence right in the face and rejects the scientific evidence for his own biased opinion. That which he knows to be impossible, is somehow true.

Once again, intelligent men have looked upon the invisible attributes of God and rejected its truth, choosing instead to believe the lie.

The Flagellum Motor

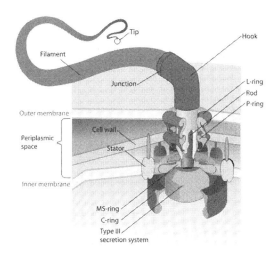

References

1. H. Halley, "The Pocket Bible Handbook" p.149
2. Peter Marshall & David Manuel, "The Light and the Glory," p.34
3. Moody Bible Institute of Science, "The Wonders of God's Creation – Human Life
4. J.K. Stevens, "Reverse Engineering of the Brain"

GOD'S UNIVERSE

The Greek astronomer and philosopher, Ptolemy, counted 1, 056 stars and claimed there couldn't possibly be more than 3,000 in the entire universe. Ptolemy's problem, of course, he hadn't read the Word of God. For had he done so, he would have understood that the stars are uncountable, they are as numerous as the sands of all the world's oceans: *"As the host of heaven cannot be numbered, nor the sands of the sea measured...."* (Jeremiah 33:22)

So how factual is this statement? Let's take a quick look. Today, aided by great telescopes such as the 200 inch telescope on Mt. Palomar in California, and the Hubble telescope mounted in orbit around the earth, we know of more than 200 billion star groups, with billions of stars in each group. Said one astronomer, *"It is a fair guess to say, that the number of stars are uncountable."*

This picture (previous page) taken from the Hubble telescope shows a tiny fraction of one area in the universe. As you can see from the picture it would be impossible to count the stars that are actually pictured here. That's because all the brighter lights are galaxies and not single stars. It is estimated there are an average of about 200 billion stars in a galaxy. If you were to go out on any given night and hold a dime at arms length, it would cover approximately the area shown in this picture. Also one should ask himself: how many stars/galaxies are behind the ones in this picture that cannot be seen?

Not only are the stars uncountable, they are receding away from the earth at thousands of miles per hour, in some cases these speeds have been measured in miles per second. So the stars you saw last night are thousands of miles further away from our planet the next night!

In 1913 an astronomer by the name of Vesto Melvin Slipher discovered that about a dozen galaxies were rapidly moving away from the earth and computed this rapid movement away from the earth to be about 700 thousand miles per hour.

It was Edwin Hubble who believed an expanding universe was the result of a tremendous explosion and this explosion blew the galaxies out away from its point of origin. Hubble postulated that galaxies were moving away from each other because of a gigantic explosion and using Slipher's figures and his own, Hubble created a new picture of the universe—an expanding universe that was the product of what today we call the Big Bang Theory.

This would be an obvious conclusion since what else could conceivably throw the universe apart in all different directions, one galaxy after another? And, if we were to reverse the recession, we could determine how long it took for the perspective galaxies to get to where they are today.

Again, it would seem that man's ingenuity and studies could bring him an answer for the origin of the universe. An answer to the question he has sought from the beginning. "How and where did this all start?"

The problem of course, is mankind doesn't like being told any of this since his scientific mind is quite able to figure it out for himself. This is called "arrogance."

Did Hubble discover the Big Bang, or is an expanding universe actually part of the creation story?

Turning the pages of our Ancient Text, let us see if it reveals to us anything about an expanding universe. After all, it's shown us everything else hasn't it?

Not surprisingly we read in Isaiah 40:22 these words: *"It is He Who sits above the circle of the earth….Who stretches out the heavens like a curtain."*

Job, the oldest book in the Bible records for us: *"He alone spreads out the heavens and treads on the waves of the sea."* (Job 9:8)

Incidentally, this verse is not only a scientific fact, it is also a prophetical fact. For Who was the only Man to ever walk on the sea? Fifteen hundred years after Job made this statement, Matthew told us what he saw:

"But the boat was now in the middle of the sea, tossed by the waves, for the wind was contrary…. And when the disciples saw Him walking on the sea, they were troubled, saying, 'It is a ghost!' And they cried out for fear. And immediately Jesus spoke to them, saying, 'Be of good cheer! It is I; do not be afraid.'

"Then those who were in the boat came and worshiped Him, saying, 'Truly You are the Son of God.'" (Matt. 14:24-33)

But that's another story, a story that can be told strictly from a prophetical standpoint, for every book of the Old Testament gives us some prophecy concerning this 'One' who walks on water and Who alone spreads out the heavens. If you were to do this study, you would find more than 300 Old Testament prophecies concerning the coming of Jesus to earth.

How many prophecies do you know that predict the coming of Mohamed or Buddha hundreds of years prior to their arrival? I know of none! There's only one, and He recorded it all for us.

But getting back to our discussion of an expanding universe, the prophet Jeremiah also described this event: *"He has made the earth by His power, He has established the earth by His wisdom, and has stretched out the heavens at His discretion."* (Jeremiah 10:12)

So, as you clearly read, Hubble didn't discover anything, it was already written in God's book thousands of years before Hubble was ever born.

Perhaps you remember the reason the Hubble telescope was put into orbit to begin with? It was to discover the end of the universe. That was one of its primary missions. When I first read this, I told a class that I was teaching at the time that Hubble would fail in this mission. The reason? My ancient text had already told me that the universe could not be measured or found out.

Our prophet Jeremiah reveals that bit of science to us, he wrote: ***"Thus says the Lord, 'If heaven above can be measured, and the foundations of the earth searched out beneath, I will also cast off all the seed of Israel for all that they have done, says the Lord."*** (Jeremiah 31:27)

The emphasis here is that He will never cast off Israel (see Isaiah 54:10), thus one can conclude that the heavens, likewise, cannot be measured.

You have read on the preceding page, that there are so many stars that we cannot possibly count them. Consider then, what a great statement David made in Psalms 147:4 when he wrote these words: ***"He counts the number of the stars; He calls them all by name."***

Back in David's day, this didn't seem like too much of a thing for anybody's god, right? But then in David's day, they only believed there to be several thousand stars at the most!

And how about the moon we see almost every night. Besides giving us a light at night (Genesis 1:16), what else does it do for us?

Well for one very important thing, the moon gives us the tides. The wave action of the tides constantly beating on the sandy beaches around the world, act as a filtering device for the oceans. Thus one could say, the moon helps to keep our oceans clean by dragging them over earth's sandy beaches. Sand, as you know, is an excellent filtering tool.

Tides are created because the Earth and the moon are attracted to each other, just like magnets are attracted to each other. The moon tries to pull at anything on the Earth to bring it closer. But, the Earth is able to hold onto everything except the water. Since the water is always moving, the Earth cannot hold onto it, and the moon is able to pull at it.

Each day, there are two high tides and two low tides. The ocean is constantly moving from high tide to low tide, and then back to high tide. There is about 12 hours and 25 minutes between the two high tides.

The gravitational attraction of the moon causes the oceans to bulge out in the direction of the moon. Another bulge occurs on the opposite side because of the conflicting gravitational field. Ocean levels fluctuate daily as the sun, moon and earth interact. As the moon travels around the earth and as they, together, travel around the sun, the combined gravitational forces cause the world's oceans to rise and fall. Since the earth is rotating while this is happening, two tides occur each day.

The term for this bulging or heaping up of the water is known as tidal heap. The seas are actually heaped up three to five feet higher under the moon and on the opposite side of the earth. This heaping on the opposite side of the earth is because of the conflicting gravitational force between the two bodies, not unlike two magnets repelling one another.

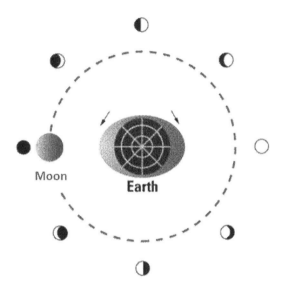

Now let's take a look and see what God's word tells us about all this: ***"Thus says the Lord Who gives the sun for a light by day, and the ordinances of the moon and the stars for a light by night, Who disturbs the sea, and its waves roar."*** And: ***"He gathers the waters of the sea together as a heap...."*** (Psalm 33:7 & Jeremiah 31:35)

The word used in Jeremiah for "ordinances" means "laws." Laws that He alone established that govern our universe. Laws that keep everything in place and doing the work that He designed it to do from the beginning. To believe all these things working together in concert as the result of a gigantic happenstance, orchestrated by an accidental explosion is pure nonsense.

WAS GOD AN EVOLUTIONIST?

Having traveled this country doing creation seminars I found that probably about half of all Christians believe in theistic evolution; that is, they believe that God used evolution to bring about His purpose. They base their opinions on the idea that evolution is a "fact." Newspapers affirm this, television programs affirm this and some of the most distinguished scientists today affirm this. Even the Pope affirms this along with many big name evangelists and pastors.

So I suppose it's not surprising that many Christians believe that if evolution is true then surely God orchestrated it.

But wait a minute, does that let Christians off the hook for their ignorance of the word? I don't think so; it is their responsibility to study God's word and theirs alone.

There was another time when the church bought into the science of the day. The science of the day was predicated on the idea that the sun revolved around the earth. This was known as the geocentric model.

Some highly esteemed scientists of the day bought into it, scientists like Ptolemy and Aristotle; Aristotle, a Greek philosopher was a student of Plato and teacher of Alexander the Great. Aristotle wrote on many subjects, including physics, metaphysics, poetry, theater, music, logic, rhetoric, politics, government, ethics, biology, and zoology. Together with Plato and Socrates (Plato's teacher), Aristotle is one of the most important founding figures in Western philosophy.

So when Aristotle spoke, people listened. But Aristotle was wrong; the sun does not go around the earth as we know today.

In 1543 Copernicus threw a wet towel over the whole thing when he published his book titled: *De revolutionibus orbium coelestium,* or *"On the Revolution of the Celestial Spheres."* In his book he predicted that the sun did not go around the earth, but rather, just the opposite. To say this caused quite a stir would be to put it mildly. The church had solidly bought into the science of the day and branded Copernicus a heretic and cast the poor lad out into the street.

How could the church brand Copernicus to be a hieratic when this idea wasn't supported in the Bible?

So here we are several hundreds of years later and today, it's the church that is blamed for teaching this false science.

Likewise, when the day comes and come it will, when evolution is overwhelmingly shown to be false, will the Christian church be blamed for promoting a false science?

I can state that churches and their constituents that uphold this false assumption of evolution are in fact denying the Word of God and His creative power; thus, to some degree they are either calling Christ a liar or they are ignorant of His Word. A very serious charge indeed!

So let me begin this discussion with a simple truth made by the Creator Himself. Jesus made this statement in Mark 10:6, **"But from the beginning of the creation He made them male and female."** Now you need to ask yourself, is this true? No slime, no monkeys, no evolution. They were simply a man and a woman from the beginning of creation!

If you reject that, then Christ cannot be who He claims to be—a liar, or was He just mistaken? Or didn't He realize that the creation story was just a myth? Wouldn't you also be interested in knowing that Jesus quotes more from the book of Genesis than any other book in the Old Testament: a book whose foundation then is based upon fables; a book that is simply story-telling, one way or another!

And what about the man and the woman, maybe it's also okay for Adam to marry Bruce since Genesis is all based on metaphors and stories anyway.

So if God needed to use evolution in order to bring forth a man then what about Lazarus? How could Jesus have possibly called him from the grave in an instant? Did that really happen? You have to wonder, huh?

You see, bringing non-life to total complete life is no problem for Him. Jesus raised Lazarus from the grave with a single command: ***"Lazarus, come forth."*** The resurrection power of His word instantly brought a dead body to full life. No evolution required here! It's also interesting to note that He specifically called the name of Lazarus, because if He hadn't, all the dead bodies in that cemetery would have been resurrected to new life! That's the power of the God of Scripture.

When the Jews were seeking a way to kill Jesus, He simply told them: ***"Destroy this temple and in three days I will raise it up."*** (John 2:19). Who would raise it up? He would of course, since He has the power to do so. Certainly there was no time for evolution to do it. If He didn't need evolution then, why would we have any reason to believe He needed it at the beginning of the creation? Or do you think that since the beginning of the creation God, over several thousands of years, learned new methods for creating people?

But the above stated problems pale compared to the very thing evolution stands for and that is death "and struggle". And, as we've just studied in this book, evolutionists have given it a name. It's called: "survival of the fittest." So let's make it very clear what God did if He used evolution. In this case, if death and struggle were in operation for millions of years prior to the arrival of human beings on the planet, then God is responsible for death, not us; because He used it in the creative processes.

This idea runs totally contrary to scripture that very plainly tell us that: ***"sin first entered the world through one man, and death through sin."*** (Romans 5:12).

Now if death reigned in the world prior to Adam through the evolution process....somebody is very, very wrong here. Either theistic evolution is wrong or God is wrong. Which do you chose, since they both cannot be right? Was there death prior to Adam? I think not!

Following closely after the idea of theistic evolution and probably as popular is the so-called "Gap Theory."

The Gap Theory is simply a way of accommodating the billions of years required by evolutionists. The idea being that the first creation i.e. ***"In the beginning...."*** God created all the life forms we now find in the fossil record. Dinosaurs, cave men, etc., but during this period Satan fell and polluted all the animals and mankind himself.

So God, disgusted with His original creation, destroyed all plant and animal life and started over. The *Scofield Bible* notes on Genesis I include the following:

"The first act refers to the dateless past, and gives scope for all the geologic ages. . . . The face of the earth bears everywhere the marks of such a catastrophe. There are not wanting intimations which connect it with a previous testing and fall of angels. . . . Relegate fossils to the primitive creation, and no conflict of science with the Genesis cosmogony remains."

It is their claim that the word *"was"* in Hebrew should be translated "became," thus rendering this reading: "And the earth became void and without form, and darkness was on the face of the deep."

One of their problems is the Hebrew verb *hayetha* (translated "was") means "was" and not: "became."

Since they believe the great flood took place during this gap-period (***"and darkness was on the face of the deep."***) Noah's flood wasn't the great flood, but rather the great flood took place prior to Adam and Eve—during the gap-period. The flood of Noah, they claim, was just a local flood.

Now if this is true, please don't tell Noah that, because he spent 120 years of his life building the ark when all he would have had to do was move!

Ok, so Noah's flood was just a local flood you say, sounds reasonable to me even if Noah had spent all that time building an ark.

Now if this is the case, what will you do with the verse where God promises Noah that He would never again destroy the earth with a flood: ***"….Nor will I again destroy every living thing as I have done."***

No more local floods???

And how about God's own statement shortly after creating the "second heaven and earth: ***"And God saw that it was good."*** Good? The very ground where Noah stood contained the remains of millions of fossils, fossils that represent sudden death and burial? That was good?

The next concoction that Christians have come up with is known as "The Day Age Theory." They speculate that a day really wasn't a day, but actually may have represented thousands, or even millions of years. Thus, they claim, it took God millions of years to create the heavens and earth. I discussed earlier in this book how ridiculous that belief is. Since God specifically uses solar time to explain each creative day. *"And the evening and* the *morning were the first day…."*

Can you imagine how confused the children of Israel would have been if they really didn't understand what a "day" really meant? And, didn't Moses know what he was talking about when he was giving the Levitical law: *"It is a sign between Me and the children of Israel forever; for in six days the Lord made the heavens and the earth, and on the seventh day He rested."* (Exodus 31:17).

Isn't it far easier, to take the text for what the writer is so plainly trying to convey?

Finally, in closing this chapter I want to reiterate the very thing the apostle Paul specifically warned the church about:

"O Timothy! Guard what was committed to your trust, avoiding the profane and vain babblings and contradictions of what is falsely called knowledge—by professing it, some have strayed concerning the faith." (I Timothy 6:20)

God does not take lightly, Christians who willfully twist His words to fit their own vain philosophies. Since the Bible is written for Christians, they, like Timothy are charged with guarding His word and using it for its express intent, and not to make it fit into a world which is at enmity with God.

"For in six days, God created the heavens and the earth, and on the seventh day He rested." Pretty plain isn't it?

THE WORLD'S GREATEST SCIENTISTS

No discussion of real science would be complete without mentioning some of the greatest scientists in the world. What we know about them and upon what they based their foundations of truth and discovery.

The reader might be surprised to know that some of the greatest scientists in the world were staunch believers in the scripture and that any theory they proposed must first and foremost align itself with the Word of God. After all, if the claims found in scripture were indeed inspired by God, then any theory advanced by them would have to be supported by scripture, otherwise it would be wrong.

While Noah isn't known for his scientific achievement, we need to consider his exploits as an engineer. After all, he built the largest boat ever constructed without the use of modern day tools. In fact, the ark was the largest boat ever constructed until the mid 1800s. And how big was the ark? Scripture is quite clear; it was 300 cubits in length, 50 cubits wide and 30 cubits high.

Obviously, the question need be posed: "How long is a cubit?" Bible scholars know that a cubit was the measurement of a man's arm, from elbow to the tip of his finger. Today, that would give us an average length of eighteen inches. This is a conservative estimate since the people of Noah's day were probably, on an average, bigger than the average person living today. However the eighteen inch cubit seems to be the favored length of most scholars so that is what I will use.

First, notice that the length of the ark was six times its width. The ancients did not build their boats using these modern day dimensions; but Noah did. I say modern day dimensions because that is exactly the same dimensions used by all modern day ships.

So was Noah just lucky choosing the size of his boat, or was he guided by a Divine source?

Next we see where God told Noah to build some ventilation for the ark. Since the heat and stench of all the animals would rise toward the top, God told Noah: *"A window shall thou make in the ark, and to a cubit shall you finish it upward."*

So we see that windows a cubit square would be built along the top (upward) portion of the boat; and would run from one end to the other, stopping a cubit from stem to stern. This would provide ample ventilation for Noah's floating zoo.

The ark was to be constructed using Gopherwood (Genesis 6:14). Gopherwood was more than likely the name given to a type of oak known to be found in abundance in that area at the time. Oak is a very closed cell wood and fairly impervious to water. In addition to this, Noah also covered the ark with tar: *"cover it inside and outside with pitch."*

We know from scripture the ark had three decks, each deck being sixteen ft. high. Since the ark was 450' X 75' we know that each deck contained at least 33,750 sq/ft of floor space. Multiply this by three, and one can see the ark contained a total floor space of 101,250 sq/ft.

Zoologists have estimated there to be around 16,000 different species of land animals during the days of Noah. Today, of course there are many, many varieties of animals within a species type. There are hundreds of differing breeds of dogs, cats, cattle, etc that didn't exist during Noah's day. Thus, if 16,000 primary species of land animals had to be loaded; and each animal took an average of five sq/ft of space[1], we are looking at a total occupied area of 80,000 sq/ft required to put one of each animal kind onto the ark. We also know that each deck was sixteen ft. high. Noah could easily stack three or more cages on top of each other. Thus quadroupling the storage capacity.

Since Noah had to load at least two of each animal kind for reprogination, we have to double the area needed for each animal kind and its mate to 160,000 sq/ft/area. But we have available more than 300,000 sq/ft of area. Keep in mind, the dimensions I have given for the ark are using the more conservative figure of eighteen inches as the standard for a cubit.

But remember, God also told Noah: *"You shall take with you seven of each of every clean animal, a male and his female;"* (Genesis 7:2)

So that makes it 14 each for every clean animal and its mate. While we are not told exactly what that means[2] it probably has more to do with solving the inbreeding problems that could occur after the animals were freed from the ark and began the process of procreation.

As I mentioned earlier in this writing, the planet had cooled considerably and during that period, most of the animals onboard the ark would hibernate.

Footnotes:

[1] *Noah wouldn't have loaded full grown animals since young, smaller animals would be healthier and more viable than full grown adults. Young elephants for example, would require about 50 sq/ft area, but a mouse, shrew, etc only 1.5sq/ft/area.*

[2] *Not to be confused with the "clean and unclean" animal kinds given by Moses for the Levitical law and ceremonial purposes. This law wasn't given by Moses until 1000 yrs after the flood. It is believed that Noah's "unclean" animals were probably a food source for other animals.*

Moses, a scientist?

Probably, outside of Jesus Himself, no other name is better known than the person Moses. And again, like Noah, Moses isn't known so much for his scientific attributes as his authorship of the first five books of the Bible known as the Pentateuch.

We marvel at the great wonders and engineering skills of the ancient Egyptians and their ability to construct magnificent artifices like the pyramids. But just as impressive is where did the engineers procure the materials from which they quarried the rock? This was an impressive feat in itself. Using only the most primitive of tools, workers quarried large stone blocks and cut it to extremely precise measurements. How did they do it? Even today, that question remains unanswered.

Moses probably knew quite a bit about what was going on during this period of history. Who knows, maybe he even had something to do with it, for scripture says of him: ***"And Moses was educated in all the wisdom of the Egyptians."*** (Acts7:22)

Moses was educated in all the wisdom of the Egyptians because he was raised in Pharaoh's house. Indeed, if Moses had stayed the course, he most likely would have become Pharaoh's successor and today, it could have been his tomb they found at the bottom of one of those pyramids. But Moses took another route, eventually rejecting everything Egypt stood for.

Why would one educated in all the wisdom of Egyptian culture; trained to become one of the most powerful and influential kings on earth, walk away from it all? The answer is quite simple, he had an encounter with the *"I AM God."* A God he never heard of or even knew existed.

But there he was, standing on top of a mountain in the middle of a howling wilderness looking at a bush that was on fire but not being consumed by the fire; a reversal, if you will, of the second law of thermodynamics.

I wonder if Moses gave this much thought; Only a Creator Who established the law, can be greater than the law and on this account, Moses realized he was standing in the presence of glory and awesomeness that cast a dark shadow forever over the ambitions of all men and their desire for greatness. For he had to know he was standing before greatness of an unimaginable magnitude and it must have been with trembling hands that he removed his sandals at the command coming from the midst of the fire: ***"Take the sandals off your feet, for the place where you stand is holy ground!"*** (Genesis 3:5)

From that very encounter, Moses became known as "Moses the man of God." One need again ask the question, how could Moses who spent forty years of his life learning Egyptian philosophy suddenly begin teaching just the opposite? Why he never incorporated any of the now discredited Egyptian superstitions into any of his writings?

In Moses' day people believed that a great man carried the earth on his shoulders and if he sneezed, it would produce a great earthquake. In Moses' day, the practice of medicine was preceded by the boiling of toads' skins and demon incantations. Yet Moses, schooled in these superstitions, wrote the most scientific treatise on modern medicine, hygiene and sanitation the world has ever known. Today in our modern scientific era, we have not added one thing or been able to improve on a single regulation he gave to the children of Israel concerning the handling of contagious diseases.

Consider for a moment highly infectious diseases. Usually if it's a particularly contagious disease, doctors will isolate the patient. It was Moses who commanded the children of Israel to move anyone with leprosy or other contagious diseases out of the camp and should they come into the camp, they were to declare themselves "unclean."

It was Moses who commanded the children to wash their hands before eating and to wash themselves and their clothing if they touched "any dead or sick thing."

It was Moses who put forth the Biblical rules for disinfection of any person who had been infected with a certain disease. Moses wrote: *"And when he that hath an issue is cleansed of his issue; then he shall number to himself seven days for his cleansing, and wash his clothes, and bathe his flesh in running water...."* (Leviticus 15:13)

Today, any doctor, before he touches a sick person, before and after will always wash his hands under the running water flowing from a tap.

It was Moses who wrote: *"Ye shall eat no manner of blood, whether it be of fowl or of beast, in any of your dwellings."* (Leviticus 7:26)

Why did he give that command? It was simple—*"For it* [the blood] *is the life of all flesh."* (Leviticus 17:11)

I heard a psychologist put forth his defense of why it doesn't really bother children to watch violence on TV. He went on to say that most people who oppose violent movies were Christians; who, he claimed, read from the Bible, "the bloodiest book in the world." He even referred to the Bible as "a book of blood."

I would like to have been able to respond to him in this manner: "I proudly plead guilty to the charge, for the only thing that gives life to our teaching and power to the Word of God is the fact that it is the blood which is the very life and power of the Gospel. The Bible declares itself to be a 'living' book, the only living book in the world. Yes sir, it is a bloody book, but life is in the blood!"

Louis Pasteur was greatly influenced by the writings of Moses and came up with the idea of sterilization. A miracle discovery of the day that saved countless numbers of lives that would have been lost, had he not learned from Moses about contamination.

Today, we know that animal fat is not good for you. Ask any doctor and he will tell you that a diet high in fat produces a high count of cholesterol. Cholesterol, they say, is a leading cause of heart attacks because of plaque build up in the arteries.

You can take this modern medical advice to the bank. Cholesterol that causes plaque build-up may kill you. Here's what Moses said about that: ***"Speak to the children of Israel, saying: 'You shall not eat any fat, of ox or sheep or goat. And the fat of a beast that dies naturally, and the fat of what is torn by wild animals, may be used in any other way; but you shall by no means eat it."*** (Leviticus 23, 24)

Moving up in time
Johannes Kepler (1571 – 1630)
Kepler was a staunch creationist. He believed that the universe must be orderly if designed by God. Kepler discovered the laws of planetary motion and conclusively demonstrated that the sun is at the center of our solar system. Kepler wrote:

"I had the intention of becoming a theologian. But now I see how God is, by my endeavors, also glorified in astronomy, for the heavens declare the glory of God."[1]

Robert Boyle (1627 – 1691)
The name of Robert Boyle is synonymous with modern day chemistry. Boyle discovered that gases consist of particles, and made discoveries concerning vacuums and is credited with inventing the first match. Boyle wrote *"The Christian Virtuoso"* to show that the study of nature was a religious duty, and in his will established the "Boyle lectures" for the proving of Christianity.

Sir Francis Bacon (1561 – 1626)
Francis Bacon was credited with developing the scientific method. Mr. Bacon wrote:

"There are two books laid before us to study, to prevent our falling into error; first, the volume of Scriptures, which reveal the will of God; then the volume of the creatures, which express His power."[2]

Bacon's writing reminds me of Job when God said to him: ***"But now ask the beasts, and they will teach you; And the birds of the air and they will tell you; Or speak to the earth and it will teach you; And the fish of the sea will explain it to you. Who among all these does not know that the hand of the Lord has done this?"*** (Job 12: 7 – 9)

The Apostle Paul also alludes to this: ***"For the wrath of God revealed from heaven against all ungodliness and unrighteousness of men who suppress the truth in unrighteousness, because what may be known of God is*** *manifest in them, for God has shown it to them. For since the creation of the world His invisible attributes are clearly seen, being understood by the things that are made."* (Romans 1:18, 19, 20)

A study of nature is thus a study of God and His creation. But today, the study of environment [nature] is an exercise in the worship of the creation rather than the Creator.

Blaise Pascal (1623 – 1662)
Blaise Pascal, a Frenchman, was a brilliant mathematician who developed the science of hydrostatics and formulated the laws of probability.

Pascal wrote: *"Except by Jesus Christ we know not what our life is, what our death is, what we are ourselves. Thus, without Scripture, which has only Jesus Christ for its object, we know nothing, and we see only obscurity and confusion in the nature of God, and in nature herself."*[3]

My favorite scientist about whom I've studied is without a doubt Sr. Isaac Newton. (1642 – 1727) Isaac Newton was probably the greatest scientist of all time. He was gifted by God with extraordinary insight into the physical laws governing the universe, or, as I referred to them in the opening chapters of this book: "God's servants."

Sir Isaac Newton said of the Bible: *"I have a fundamental belief in the Bible as the Word of God, written by men who were inspired by His Spirit. I study the Bible daily"*.[4]

But Newton went further than this by saying: *"All my discoveries have been made in answer to prayer."*[5]

Carolus Linnaus (1707 – 1778)
Linnaus laid the foundations for modern day taxonomy which is still called the Linnaean system. Linnaus wrote in his writings titled *Systema natura*, these words:

"I saw the infinite, all-knowing and all powerful God....I followed His footsteps over nature's fields and saw everywhere an eternal wisdom and power, an inscrutable perfection."[6]

Samuel Morse (1791 1872)
You got it. Samuel Morse invented the telegraph and Morse Code; built the first camera and founded the National Academy of Design. Interestingly enough, Morse also established one of America's first Sunday schools. Morse wrote:

"The nearer I approach to the end of my pilgrimage, the clearer is the evidence of the divine origin of the Bible, the grandeur and sublimity of God's remedy for fallen man are more appreciated, and the future is illumined with hope and joy."[7]

Michael Faraday (1791 – 1867)
Faraday developed the science of electricity and electromagnetism, invented the generator and transformer. He also pioneered the liquefaction of gases and discovered benzene. Faraday wrote:

"The Bible and it alone, with nothing added to it nor taken away from it by man, is the sole and sufficient guide for each individual, at all times and in all circumstances....Faith in the Divinity and work of Christ is the gift of God, the evidence of this faith is obedience to the commandment of Christ."[8]

James Joule (1818 – 1889)
It was James Joule who proved the mathematical relationship of electrical energy and heat loss. It was through his research he was able to prove the law of energy conservation (Ecclesiastes 3:14) and today, the term Joule is still used to describe a unit of energy. Joule said:

"It is evident that an acquaintance with natural laws means no less than an acquaintance with the mind of God, therein expressed in the Scripture."[9]

John F. Herschel (1738 – 1822), the son of noted astronomer Sir William Herschel, a great astronomer in his own right and avid creationist; John discovered more than 500 stars and nebulae, declared:

"All human discoveries seem to be made only for the purpose of confirming more and more strongly the truths that come from on high and are contained in the sacred writings."[10]

Joseph Lister (1827 – 1912). It was Joseph Lister who was most responsible in developing antiseptic surgery through the use of disinfectants. In fact, the popular antiseptic mouth wash used today (Listerine) was named after him. Lister, a devout Christian was knighted and made president of the Royal Society. He also became president of the British Association for the Advancement of Science. Lister declared:

"I have no hesitation in saying that in my opinion there is no antagonism between the Religion of Jesus Christ and any fact scientifically established."[11]

Dr. Walter von Braun (1912 – 1977), what can one say about this great rocket scientist? Captured during the Second World War, von Braun was later brought to the United States because of his tremendous knowledge of rocketry. Walter von Braun later would become the director of NASA's Space Flight Center and was most responsible for the design of rockets that eventually landed man on the moon. Von Braun declared:

"There are those who will argue that the universe evolved out of randomness. But what random process could produce the brain of man or the system of the human eye?"[12]

Von Braun unashamedly and on a regular basis prayed for the safety of those on the manned missions he planned.

If you remember, it was Christmas Eve, 1968 when Apollo 8, first went into lunar orbit. On that evening, Commander William Anders sent this message back to earth:

"We are now approaching lunar sunrise and, for all the people back on Earth, the crew of Apollo 8 has a message that we would like to send to you."

"In the beginning God created the heaven and the earth. And the earth was without form, and void; and darkness was upon the face of the deep. And the Spirit of God moved upon the face of the waters. And God said, Let there be light: and there was light. And God saw the light, that it was good: and God divided the light from the darkness."

Anders crewmate, Jim Lovell continued reading from the Word of God:

"And God called the light Day, and the darkness he called Night. And the evening and the morning were the first day. And God said, Let there be a firmament in the midst of the waters, and let it divide the waters from the waters. And God made the firmament, and divided the waters which were under the firmament from the waters which were above the firmament: and it was so. And God called the firmament Heaven. And the evening and the morning were the second day."

Frank Borman finished the reading:

"And God said, Let the waters under the heavens be gathered together unto one place, and let the dry land appear: and it was so. And God called

the dry land Earth; and the gathering together of the waters called he Seas: and God saw that it was good."

"And from the crew of Apollo 8, we close with good night, good luck, a Merry Christmas – and God bless all of you, all of you on the good Earth."

It was an extremely reverent moment. I watched in awe as the camera aboard Apollo 8 was shut off. In Mission Control, silence prevailed for a bit as the residents of *"the good earth,"* considered what they had just heard from lunar orbit.

Not just great explorers, but great scientists in their own right, reading scripture from the greatest book ever written; were conveying their very thoughts to all the residents of earth. It was indeed an extraordinary moment for all mankind. The gift God has given to us all, His Holy Word, broadcast over the entire planet.

Noted atheist, Madalyn Murray O'Hair was so enraged by it all; she promptly tried to sue the US Government alleging violation of her first amendment rights. However the suit was dismissed by the US Supreme Court due to lack of jurisdiction. The poor thing was found years later, hacked to death and buried in a shallow grave.

In 1986, O'Hair wrote an essay for the American Atheist about her hopes that nothing special would happen to her body. She didn't want any *"dirty Christers to get their hands on her corpse."*

"A dead body", O'Hair wrote, *"was nothing more than a fallen leaf from a tree, a dog killed on the highway, a fish caught in a net."*

Madalyn O'Hair was murdered along with several of her sons. Their hacked up and burned bodies were found in a shallow grave on a Texas ranch. Later Her estranged son William, still a practicing Baptist Minister, interred Madalyn, Jon, and Robin into a common vault with an unmarked grave, with no religious fanfare.

It appears that God granted Ms O'Hair her final wish.

I could continue on and on, as the list of great scientists who were avid Bible believers goes on and on. But you get the picture. Not only scientists, but also some of the greatest statesmen in the world were indeed Christians who studied their Bibles on a daily basis. And need I even mention our own founding fathers? Signers of probably the greatest document ever devised by man himself outside of the Bible—The US Constitution.

It was at the first Constitutional Convention of 1787, that James Madison proposed the plan to divide the central government into three branches. He discovered this model of government from the Perfect Governor, as he read *Isaiah 33:22:*

"For the LORD is our judge, the LORD is our lawgiver, the LORD is our king; He will save us."

And as you have just read, from this verse we have in place today a government made up of the Judicial, Legislative and Executive branches of our government!

Atheists today are trying to revise our history books so that young Americans might never discover that this nation was founded on the Word of God by men of honor. So completely sure were these men that God's Word was absolute truth; they led the fledgling country in their footsteps.

So absolutely entwined in Scripture are the laws which gave rise to our great nation that other people in other countries could only marvel at what they were seeing. Indeed, many would come to see for themselves a country whose people were being led by the Word of God; a people who rejoiced in the many freedoms they found therein.

"You shall know the truth, and the truth will set you free" seemed to be the American motto. So impressed with this type of love for, and reverence of His Holy Word, a French Nobleman by the name of Alexis De Tocquville, after traveling America in 1835 would write of his findings:

"...there is no country in the world in which the Christian religion retains a greater influence over the souls of men than in America; and there can be no greater proof of its utility, and of its conformity to human nature, than that its influence is most powerfully felt over the most enlightened and free nation on earth.

"Americans combine the notions of Christianity and of liberty so intimately in their minds, that it is impossible to make them conceive the one without the other."

One of America's great generals, Wesley Merritt, superintendent of the U.S. Military Academy at West Point spoke these words in 1882:

"The principles of life as taught in the Bible, the inspired Word, and exemplified in the matchless life of Him 'who spake as never man spake,' are the rules of moral action which have resulted in civilizing the world."

Yes, America became great because she followed after the "Great God" of all creation. It's time America, to turn back.

References

1. John H. Tiner, "Johannes Kepler, Giant of Faith and Science," 1977, p. i
2. H.Morris, "Men of Science," p15
3. Blaise Pascal, "Thoughts, Letters, and Opuscules," 1875, p335
4. John Hudson Tiner, "Isaac Newton: Inventor, Scientist, and Teacher," 1975, p. i
5. D.C.C. Watson, "Myths and Miracles: AvNew Approach to Genesis," 1988, p.112
6. Charles Gillispie, "Dictionary of Scientific Biography," Vol. 8, 1973, p.380
7. H. Morris, "Men of Science, Men of
8. Ibid, p.37,38
9. J.C. Crowther, "Men of Science," 1936, p.139
10. Tyron Edwards, Comp., "The New Dictionary of Thoughts: An Encyclopedia of Quotations," 1957, p.49
11. Rhoda Truax, "Joseph Lister: "Father of Modern Surgery," 1944, p.121
12. Wernher von Braun, in a letter read by Dr. John Ford to California State School Board of Education. Sept. 14,1972

Notes

INDEX

A

Abortion, 16, 132, 133
Actinic radiation, 91
Adaptation, 53, 55
Adiabatic Lapse Rate, 83, 87
Alaska, 111, 112, 122, 134
Alaska Kodiak, 54, 89
Alaskan muck, 109, 110, 111
Allbrook, David, Prof. 16
Alvin (submarine), 148
Amazon (rain forest), 130, 131
Ambient air temp. 83
Amino acid sequence, 20
Antarctic (glaciers), 121, 126
Antediluvian age, 92
Anthropoidea, 47
Antiparticle, 72
Ape, 47, 49, 50, 51
Ape-man fossils, 93
Appendix (human), 27
Apollo 8, 182
Ararat mountains, 119, 145
Arctic tundra, 109
Ardry, Robert, Prof. 55
Aristotle, 167
Arizona, 125
Ark (Noah), 10, 43, 90, 99, 113, 119, 143, 145, 170, 173, 174, 175
Ark of the Covenant, 143
Atmosphere (layered), 81, 122
Atmosphere rich in O_2, 90
Atmosphere O_2 free, 21
Atmosphere rich in CO_2, 90
Atmospheric ozone, 91
Atmospheric pressure, 90, 96
Atoms, 65, 66, 67, 69, 71, 72, 73, 74, 123
Atomic clock, 64
Atomic theory, 66
Atoms radiate light, 72
Australia, 16, 120
Australopithecus, 50, 51, 116
Axelrod, Prof. 42

B

Baboon fossils, 116
Background radiation, 68
Bacon, Sir Francis, 178
Bacterial flagellum, 157, 158
Balance of clouds, 144, 145
Baugh, Carl, 79, 94, 95
Beadle, George, 91
Bean pods, 109
Bears, fossil, 54, 89, 110, 112
Beaver, fossil, 89, 92
Bedrock, 109
Beresovka mammoth, 111
Big Bang neutrinos, 67, 68
Big Bang theory (sick pall), 18
Bio-organic molecules, 24
Biogenetic law, 28
Bird, Windell, 21, 31
Bison, fossils, 110, 111
Biston betularia (Peppered moth), 10

Blaise, Pascal, 179
Bock, W., 28
Bonaparte, Napoleon, 43
Bongaarts, John, 135
Borman, Frank, (astronaut), 181
Bowden, Martin, 49
Boyle, Robert, 178
Brady, R., 53
British Museum of Nat. History, 12, 16, 55
Buddhism, 59
Burbidge, G., 19
Butter-cup flowers, 111

C

California's Mt. Palomar, 161
Cambrian formation, 42
Carbon dioxide, 7, 90, 97, 124, 127, 128, 131
Catastrophic events, 108, 116, 127, 132
Cave man, 27, 47
Celestial bodies, 18, 168
Chaos cannot produce order, 8, 73
Charged particles, 66, 69
Chimpanzee, 76
Chlorofluorocarbons, 121
Christ Jesus, 8, 9, 10, 31, 61, 62, 68, 69, 70, 142, 152, 163, 168, 169, 175, 179, 180
Circuit of the wind, 146
circular reasoning, 32
Civilization, 133
Clark, Austin, 43
Climate change killed dinosaurs, 96, 97, 116, 118
Climate Model, 126
Climate, 7, 97, 117, 118, 126, 127
Coacervates, 20
Cold Front, 82, 83, 84, 105
Columbia Valley, 109
Columbus, Christopher, 149
Comets, 18, 97, 116, 147
Comets Killed Dinosaurs, 97

Communism, 15, 35
Conception initiates life, 154
Conducting gas, 144
Conifer Forests, 80
Conservation of energy, 13, 17
Continental drift, 103
Continents, 119, 131
Cook, Harold 48
Copernicus, 168
Cordaites, 89
Coulee, Grand, 109
Coulee, Moses, 109
Cray supercomputer, 153
Creation Science, 4, 9, 12, 25, 49, 59, 71, 73, 80, 94, 100
Cro-Magnon, 50
Cubit, (Biblical measurement), 173

D

Dalton, John, 66
Darwin, Charles, 15, 20, 24, 35, 153
Darwin's warm little pond, 20
Das Kapital, 15, 35
Dawkins, Richard, 42, 157
deSitter, Willem, 62
Death star killed dinosaurs, 116
Denton, Michael, 20, 76
Deserts are expanding, 133
Deserts will shrink, 125,
De Tocquivlle, (French Nobleman), 183
Diatoms, 131
Dillow, Joseph, 81, 90, 111
Dinosaur tracks, 81, 82, 94, 95
Dinosaur (s), 81, 82, 93, 94, 95, 96, 97, 118
Dinosaur's lung capacity, 96
DNA, 151, 152
donkey, 75, 76
Dubois, Eugene, 49
Dunbar, C.O. 35
Dust of the earth, 150
Dust Particles, 82, 87, 104, 105, 116, 117, 147

E

Each after its own kind, 75, 76
Ear muscles, 27
Earth brought forth life, 75
Earth, circle, 149, 163
Earth covered by a flood, 94, 99, 100, 101, 102, 103, 105, 109, 112, 117, 118, 120, 146, 170
Earth created to be inhabited, 121, 123
Earth hung on nothing, 11
Earth, like a garden, 57, 87, 88, 105, 112
Earth rotation, 10
Earth stretched out over water, 101, 102, 103
Earth divided, 119
Earth, was sphere of water, 73, 74
Earth without form, void, 62, 73, 162, 181
Earth, atmosphere, 80, 125, 127, 147
Earth's crust, 31, 91, 100, 101, 102, 103, 117, 118
Earth's forests being destroyed, 123, 127, 128, 129, 130, 133, 135
Earth's foundations, 118, 164
Earth's molten core, 74
Earth's topographical features, 83
Earthquakes, 102, 133
Eastern mysticism, 57
Eccentric orbit, 18
Egypt, 141, 172, 176
Ehrlich, Paul, 22
Einstein, A. 62, 63, 157
Electric charge, 144
Electromagnetic force, 69
Electron, 67, 69, 72
Elephant, 10, 54, 89, 111
Embryo, human, 28, 29, 154
Embryo's, recapitulating, 28
Energy conversion, 11
Energy, light exchange, 67, 69, 72
Entropy, 2
Environmentalist, 121, 133, 134, 135
Equator, 113, 115, 119

Equatorial regions, 113, 119
Erebus, Mt. 122
European settlers, 131
Evaporating pools, 20
Everest, Mt. 99, 100, 118
Evolution is anti knowledge, 16
Evolution, chemical, 20
Evolution is faith, 12, 16, 45, 49, 55, 57, 58, 79
Evolution, organic, 16
Evolution, phylogenetic tree, 12, 55
Expanding universe, 162, 163
Explosion, Big-Bang, 17, 18, 19, 64, 67, 68, 162
Explosions produce order? 17

F

Faraday, Michael, 180
Flew, Antony, 151
Famine, millions will starve, 132, 133, 135
Father of lights, 72, 73
Ferdinand, King, 149
Firmament, 74, 181
Fish, Kent, Prof., 151
Fittest didn't survive, 24, 54, 55
Flood epoch, 2, 104, 119, 120
Floods stood upright, 142
Fontaine, Maury, 148
Forest fires, 4, 132
Forests are to thick, 131
Fossil beds, 116
Fossil, index, 32
Fossil, polystrate, 43, 44
Fossil record, 35, 36, 41, 42, 43, 45, 54, 55, 88, 89, 90, 93, 99, 112, 169
Fossil species appear fully formed, 42, 45
Fossil, transitional, 36, 43
Fossilization, 115
Fossil tracks, 81, 82, 94, 95
Fossil, baboon, 116
Fossils, Cambrian, 42
Fossils, coelenterates, 42

Fountains of earth, 100, 101
Freon, gas, 121
Friedmann, Alexander, 62
Frontal lifting, 83, 84
Fruit flies, 25

G

Gaia hypothesis, 59
Gale force winds, 108
Geike, Archibald, 108
Geike's Maxim, 108
Genes, 54, 77, 151, 152
Genetic code, 23
Genetic improvement, 25
Geologic column, 31, 32, 33, 44
Geologic uniformity, 32
Gibbon, 49
Gilgamish Epoch of the flood, 10
Gish, Duane, Prof. 15, 19, 20, 49, 151
Glacial ice, 110, 115, 119, 120
Glaciation, 108
Global cooling, 116
Global education, 58
Global warming, 6, 126, 127, 130
God's Invisible Attributes, 73, 156, 157, 158, 175, 179
Gore, Al, 127, 128, 130
Gorillas, 45
Gould, Steven, 18, 32, 36, 45
Gow, Anthony, 109
Grand Canyon, 152
Grasse, Pierre, 23, 55
Gravitational fields/attraction, 62, 64, 65, 147, 164
Gravitational time dilation, 64, 65
Gray, William, Dr. 127
Greenhouse effect, 123, 126, 127
Greenwich, England, 64
Growth rings, 89

H

Haeckel, E., 28
Hail, 109, 112

Hairy mammoth, 54
Hammer, fossil, 95
Harris, Sidney, 20
Heavens spread out by God, 11, 163
Helium, 18, 19
Hindu bible, 10
Herschel, John, 180
Hesperopithecus, Nebraska man, 4822, 49
Hibernate, 113, 175
Hinderliter, Prof. 79
Hinduism, 59
Hoary frost, 109
Homo erectus, 35, 51
Homo sapien, 50
Homologous structure, 76
Horizontal variations, 23
Horse, 75, 76
Host of heaven, 161
Hoyle, Fred, 17, 18
Hull, D. 20
Human artifacts, 82, 95
Human body, 151
Human body extremely complex, 151
Human brain, 23, 134, 152, 153, 181
Human embryo, 28, 29, 154
Humanism is a religion, 57, 58
Humphery's, Russell, Prof. 64
Hutton, James, 31
Huxley, Julian, 57
Hydraulic force of water, 115
Hydrochloric acid, 122
Hydrogen, 17, 18, 19, 67, 150
Hydrogen atoms, 67
Hyperbaric oxygenation, 90
Hyperons, 67

I

Ice age, 105, 108, 109, 110, 112, 119, 133
Ice caves, 109, 110
Ice sheets, 105, 113, 119, 120
Idaho, 109

Idso, Sherwood, Dr. 125
Indonesia, 120
Insecticides, adaptation, 54
Insectivores, 47
Ionized pathway, 144
Ionosphere, 145
Isabella, Queen, 149
Israel, 4, 141, 142, 143, 164, 171, 176, 178

J

Jastrow, Robert, 18
Java man, Pithecanthropus, alalus, 49
Jericho, 9, 10, 143
Jesus is "Light of the world," 168
Jet stream, 122
Johanson, Donald, 51, 93
Johnson, Philip, Prof. 43
Jordan river, 143
Jupiter, 18, 147, 148
Jeremiah, prophet, 69, 155, 161, 163, 164, 165
Job, book of, 10, 11, 68, 96, 101, 104, 112, 119, 142, 143, 144, 145, 148, 150, 154, 155, 163, 178
Joule, James, 180
Jovian surface, 147

K

Kettles, 44
Kepler, Johnnes, 178
Kinds always produce like kinds, 75
Kuban, Glen, 94

L

Laminin molecule, 156
Lava rock, 109, 110
Lazarus, 168, 169
Leaky, Richard, 93
Life appeared suddenly, 42, 45
Light energy, 63, 65, 67, 72

Light energy and mass are equal, 63, 64, 72
Light holds universe together, 63, 65
Light is refracted, 145
Light particles, 67
Light quantum, 67
Light spectrum, 145
Light year, 74
Lightning, 142, 143, 144, 145
Linnaus, Carolus, 179
Lion, 76
Lister, Joseph, 180
Local flood theory, 170
Logging, 128, 131
Longevity, 90, 91
Longwave radiation, 92
Los Alamos National Labs, 144
Low pressure weather systems, 147
Lucy, Australopithecus Africanus, 50, 51
Lyell, Charles, 108

M

Macroevolution, 16
Magmatic water, 101
Magnesium chloride, 67
Magnetic energy, 63
Mammoth, 54, 107, 108, 110, 111
Man, pre-historic, 48
Mars, 17, 149
Marx, Karl, 15, 35
Mason, Brian, 81
Mastodon, 110
Matter, energy, 11, 12, 16, 17
Matthews, L. 24
Mercury, 17, 148
Merritt, Wesley, 184
Mesozoic period, 117
Meteorites, 81, 97, 147
Meteorological events, 82
Methuselah, 93
Microevolution, 16
Mid-Atlantic Ridge, 101, 102, 106
Moist air, lighter than dry air, 83, 84
Molecular make-up of universe, 17

Molecular substance, 73, 74
Molecules, 16, 19, 20, 66, 67, 73, 90, 123, 155
Monkeys, 47, 166, 168
Moody Institute of Science, 153
Moon, 18, 80, 147, 148, 152, 164, 165, 181
More, Louis, Prof. 58
Morris, Henry, 25, 27, 44, 55
Morris, John, 44
Morse, Samuel, 179
Moses, 73, 109, 142, 171, 175, 176, 177, 178
Mother earth, 59
Mountain masses/ranges, 103, 117, 118
Mountains covered 20' by water, 100
Mountains of Ararat, 119, 145
Mountains pushed up, 103, 117, 118
Mutations, 23, 24, 25

N

Naked Ape, Ramapithecus, 49
NASA, 18, 121, 123, 126
National Geographic, 49, 51, 52, 93
Natural Selection, 24, 25, 53, 54, 55, 152, 153
Neanderthal man, HomoNeanderthalensis, 24
Nebraska man, Hesperopithecus, 48
Neutrinos, 67, 68
Neutrons, 66, 67, 69
New Age Movement, 57, 58, 59
New Mexico, 61, 144
Newton, Isaac, 18, 65, 179
Newton's law of gravity, 65
Nitrogen, 19, 150
NOAA, 122
Noah, 10, 43, 90, 92, 100, 101, 113, 120, 145, 170, 173, 174, 175
Nuclear force, 67
Nuclei for condensation, 82, 83, 87, 104, 105
Nucleus of atom, 67, 69, 71, 72, 90
Nutcracker man, Zinjaanthropus bosi, 51

O

O'Rourke, 32
Ocean floor, 102
Olduvai Gorge, 51
Orangutans, 45
Ordinances, (God's Servants), 141, 142, 165
Order of the Universe, 18
Oregon, 109
Oreographic lifting, 83
Organic carbon, 131
Organs, vestigial, 27, 28
Origin of life, 22
Over-population myth, 123, 133
Oxygen atom, 90
Oxygen molecules, 90, 123
Ozone, 20, 91, 92, 121, 122, 135
Ozone canopy, 91, 92
Ozone hole, 122, 135
Ozone myth, 121

P

Pair annihilation, 72
Palm trees, 112
Paluxy river, 81, 94, 95
Particle spin, 67, 72
Pasteur, Louis, 21, 177
Pathfinder of the sea, 148
Pathway for lightning, 143, 144
Patterson, Colin, 12, 16, 55
Peking man, Sinanthropus Pekinensis, 49
Peppered moth, 24, 53
Permafrost, 108, 110
Petition (Global warming), 127, 128
Petrifaction, 115
Phosphoric salts, 20
Photon(s), 67, 69
Photosynthetic phytoplankton, 131
Pi mesons, 67, 69
Piltdown man, Eoanthropus Dawsoni, 49
Pinatubo, Mt., 122
Pineal gland, 27

Planets, 18, 64, 65, 147
Plasma gas, 144
Plato, 167
Polar ice, 109, 120, 123, 126
Polynucleotide's, 20
Polypeptides, 20
Polysaccharides, 20
Popper, Karl, 53
Population Bomb – Book, 14, 132
Pre-flood world/strata, 112
Pre-formation theory, 154
Prehistoric, 35, 48
Present is key to the past, 31, 32, 103, 110
Primeval earth, 19, 73, 74
Primordial sea, 20, 21
Proteins, 20, 155
Protons, 35,36,37, 156, 157
Provine, Prof. 59
Prudhoe bay, 112
Ptolemy, Greek astronomer, 161, 167
Punn, P. Prof. 20

R

radiation shortens lifespan, 91
radiation sickness, 91
Radio waves, 63, 68
Rain, 64, 82, 83, 87, 100, 102, 104, 108, 113, 116, 135, 136, 137, 138, 145, 146
Rain forest, 128, 129, 130, 131
Random mutations, 23
Recapitulating embryos, 28
Red Sea, 142
Relativity, Theory, 61, 62, 74, 75
Rhode Island, 134,
Rifkin, Jeremy, 58
Rimmer, H. 28, 51, 66, 75
Rock stratum, 44, 51, 81,
Russia, 4, 35, 81

S

Sagan, Carl, 50, 130
Sanderson, Ivan, 110

Satellite, 121, 125, 127, 130, 146
Satellite, (Aqua), 126
Satellite, (Shoemaker Levy), 147
Satellite, (Tiros), 125
Saturation point, 83
Saturn, 18
Schneider, Stephen, 126
Scopes trial, 49
SD School of Mines & Technology, 5, 121
Sea floor has sunk, 68
Sea shut in with doors, 118
Seamounts, 118
Seas rose from melting ice, 119
Sedges, 107
Sedimentary rock, 104, 105, 115
Sedimentation, rapid rate of, 44
Shapiro, R. Prof. 21, 122, 123
Siberia, 107, 110, 111
Simak, Clifford, 88
Simpson, George, 42
Singularity, 17, 19
Slipher, Melvin, 162
Smoke particulates, 82, 104, 105, 117
Solar radiation, 92
Solar system, 18, 64, 147, 168
Solar wind, 90, 92
Solomon, J. 68
South pole, 101, 121, 123
Spenser, Roy, Dr. 127
Spontaneous generation, 21, 22, 158
Springs of the sea, 148
St. Augustine, Mt., 122
Star groups, 161
Stars, 18, 63, 80, 161, 162, 164, 165, 180
Stars uncountable, 161, 164
Stormy weather, 87
Strata, pre-flood, 81
Stratigraphic column, 31
Stratosphere, 81, 122
Strickburger, Monroe, 28
Strong nuclear force, 67
Subatomic, 61, 67, 71
Sub-human, 27, 50

Substance yet unperfected or unformed, 154
Subterranean water source, 100
Sukachev, Victor, 111
Sun, glory of, 18
Sun is shrinking, 63, 64
Sunshine, 117, 145
Sunlight weighs 3 lbs. 64

T

Tautology, (circular reasoning), 24, 53
Tannin oak, 137
Tektites, 81
Telescope, (Hubble), 148, 161, 162
Terrarium, 86, 87, 123
Terrestrial bodies, 18
Tertullian, 133
Theistic evolution, 15, 169
Thermal blanket (atmosphere), 105
Thermal convection, 82
Thermodynamics (1st law), 11, 12, 18, 120
Thermodynamics (2nd law), 11, 12, 17, 18, 20, 25, 71
Thermodynamics (chemical), 20
Thick darkness (smoke, dust), 117
Third world, 132, 135
Thomson, Dietrick, 19
Thunder cell, 83
Thunderbolt (lightning), 143
Thyroid gland, 27
Tides, tidal heap, 165
Time Magazine, 94, 95, 116
Tiros weather satellite, 145
Tonsils, 27, 28
Transitional advancement, 43
Treaty, biodiversity, 130
Tree ferns, 88
Tree growth rings, 89
Trees produce oxygen, 131

U

Ultraviolet light, 91, 123
Ultraviolet radiation, 90, 116
Uniformitarianism, 13, 14, 122
Universe has order, 18, 19
Universe wants to be known, 19

V

Valence, 67, 74
Vapor canopy, 74, 82, 88, 89, 90, 91, 92, 96, 103, 105, 109
Vapor (water), 74, 75, 80, 81, 82, 83, 90, 92, 104, 105, 108, 124
Venus, 17
Vestigial organs, 27, 28
Volcanic activity, 109, 117
Volcanic ash, 105, 109, 116, 117
Volcano's killed dinosaurs, 116, 118
Von Braun, Walter, 181

W

Washington (state), 62
Water cluster ions, 45
Weather system, 71, 85
Weight for wind, 83
West Point, 18
Whale skeleton, 11
White Sands Proving Grounds, 32
Williams, Lindsey, 64
Windows of heaven, 1, 57, 60
Wolves, 27, 63
World government, 30
World humanism, 29, 30
World is overpopulated, 77
World religious system, 30
World Resource Institute, 79
World will run out of gold, 77